STUDENT UNIT GUIDE

UNIT

OCR (B) AS F331

Salters

Chemistry

Chemistry for Life

Frank Harriss

To Maggi

Philip Allan Updates, an imprint of Hodder Education, part of Hachette Livre UK, Market Place, Deddington, Oxfordshire OX15 0SE

Orders
Bookpoint Ltd, 130 Milton Park, Abingdon, Oxfordshire OX14 4SB
tel: 01235 827720
fax: 01235 400454
e-mail: uk.orders@bookpoint.co.uk
Lines are open 9.00 a.m.–5.00 p.m., Monday to Saturday, with a 24-hour message answering service. You can also order through the Philip Allan Updates website: www.philipallan.co.uk

ISBN 978-0-340-94821-7

First printed 2008
Impression number 5 4 3 2 1
Year 2013 2012 2011 2010 2009 2008

Typeset by Fakenham Photosetting, Norfolk
Printed by MPG Books, Bodmin

P01216

Contents

Introduction

■ ■ ■

Content Guidance

■ ■ ■

Questions and Answers

Introduction

About this guide

This guide is designed to help you prepare for the first Salters AS unit test, **Unit F331: Chemistry for Life**. This unit is divided into two sections: **The elements of life** and **Developing fuels**.

The aim of this guide is to provide you with a clear understanding of the requirements of the unit and to advise you on how best to meet those requirements.

The book is divided into the following sections:

- This **Introduction**, which outlines revision and examination technique, showing you how to prepare for the unit test.
- **Content Guidance**, which provides a summary of all the 'chemical ideas' in Unit F331.
- **Questions and Answers**, in which you will find questions in the same style as in the unit test, followed by the answers of two students, one of whom is likely to get an A grade, the other a C/D grade. These answers are followed by examiner's comments.

How to use this guide

- Read the section 'Revision and examination technique' in this Introduction.
- Decide on the amount of time you have available for chemistry revision.
- Allocate suitable amounts of time to:
  - each section of the Content Guidance, giving the most time to the areas that seem most unfamiliar
  - the questions from the Question and Answer section
- Draw up a revision timetable, allocating the time for questions later in your timetable.
- When revising sections of the Content Guidance:
  - read the guidance and look at corresponding sections in your notes and textbooks
  - write your own revision notes
  - practise answering questions from past unit tests and from other sources, such as *Chemical Ideas*.
- When using the Questions and Answers:
  - try to answer the question yourself
  - then look at the students' answers, together with your own, and try to work out the best answer
  - then look at the examiner's comments

Revision and examination technique

How do I find what to learn?

In addition to this guide, other useful sources are:

- the specification. This is the definitive one. If it's not in the specification it won't be in the paper. However, the wording in the specification is written in 'examiner-speak', so it might not always be absolutely clear what is required. These notes should help you to interpret the module content — every specification point is covered in the Content Guidance section.
- the 'Check your knowledge and understanding' activities in the *Activities* pack. These also suggest sources of details not found in the *Chemical Ideas* book. Some of the material is in the *Storylines* book and some in the activity sheets.
- your own and your teacher's notes. Preparation for an exam is not just something you do shortly before you take the paper. It should be an integral part of your daily work in chemistry. If you've left it a bit late this time, remember this when you are preparing for later units.

How much of the Storylines and Activities do I need to learn?

Have a look through for yourself, but you will find the details in the 'Check your knowledge and understanding' activity sheets referred to above.

The primary function of the *Storylines* book is to provide a framework and a justification for studying the theory topics. In the case of **The elements of life**, most of the theory is in *Chemical Ideas*. However, there is a lot more theory in the storyline of **Developing fuels**.

The activity sheets are provided to teach practical and other skills and to back up the theoretical ideas. However, they too contain some theory that may occur elsewhere.

General revision tips

Revision is a personal thing

What works for one person, does not necessarily work for another. You should by now know what methods suit you, but here are a few ways to set out your revision notes.

- Mind maps — ideas radiate out from a central point and are linked together; some people like to colour these in.
- Notes with bullet points and headings.
- Small cards with a limited 'bite-size' amount of material on each.

Make a plan

Divide up your material into sections (the Content Guidance section will be helpful here). Then:

- work out how much time you have available before the exam

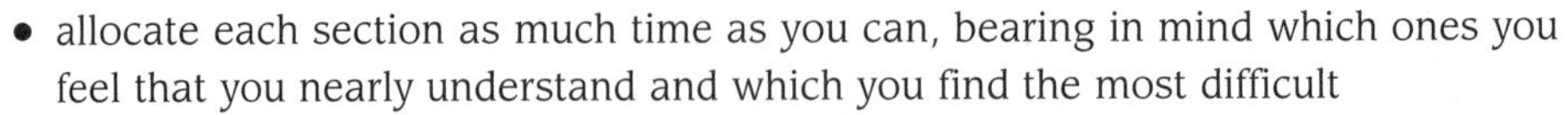

- allocate each section as much time as you can, bearing in mind which ones you feel that you nearly understand and which you find the most difficult
- fit this in with any revision your teacher is going to do — ask him or her for a summary

Write, write, write

Whatever you do, make sure that your revision is *active*, not just flipping over the pages thinking that you know it already. Write more revision notes, test yourself (or each other), and *try questions*.

Test yourself

- The questions in *Chemical Ideas* are useful 'drill exercises' on topics, but are not all like exam questions.
- If you have taken end-of-module tests, go through them again and then check your answers against your corrections or the model answers you may have been given. These are much more like exam questions.
- Past papers. For the new OCR specification, there are specimen assessment materials available on the OCR website. As time goes by, there will be more and more past papers available. Many of the past papers (for Module 2850) contain valid questions too, and your teacher should be able to advise you.
- The Question and Answer section of this book is designed specifically to allow you to test yourself.

Know the enemy — the exam paper

I hope, since you will have prepared properly, you will look on the exam as an opportunity to show what you can do, rather than as a battle. Be aware, however, that you must prepare yourself for an exam just as you would for an important sporting contest — be focused. Work hard right through the 75 minutes and do not dwell on difficulties — put them behind you. Try to emerge feeling worn out but happy that you have done your very best, even if you have found it difficult (others will probably feel the same way). Then forget it and don't have a post-mortem.

Every question tells a story

Salters is all about learning chemistry in relevant (i.e. real-life) contexts, so it is right that the exam papers should reflect this. Sometimes the context comes from *Storylines*; sometimes it is a new one. Look carefully at the 'stem' (the introduction at the top of the question). Most of the important facts here will be needed somewhere in the question. Sometimes, small, additional stems are added later on. These are important too.

Sixty marks in 75 minutes

If you have trouble getting through exam papers on time (or if you tend to rush), plan to pace yourself through the paper, so you can tell whether you are ahead of or behind the clock. There is a grid on the front of the paper that gives the marks for each question, which will be helpful here. It is best to work through the paper, from question 1 to question 4, since the first question is intended to be one of the easier ones.

50/50 knowledge/application of knowledge

This one you may not know about. About 30 of the marks test your knowledge and ask about things you will have learned. The other 30 marks are for the application of that knowledge to new situations or through doing calculations. These questions often begin 'Suggest...' to make it clear you are not expected to be able to recall the answer. However, you can leave the balance to the examiner. There are about the same number of marks on the chemical ideas from **The elements of life** as there are on the chemical ideas from **Developing fuels**.

Easy and hard parts

The papers are designed so that, ideally, an A-grade candidate will get 80% (48/60) and an E-grade candidate 40% (24/60). The actual mark for each grade varies between papers, depending on the difficulty, and is only decided after all the papers have been marked. Some of the marks are designed with A-grade candidates in mind, and thus will seem quite demanding. Other marks are designed in order to allow an E-grade candidate to score 40%, and so will seem rather easy. Thus, there will be easy, middling and hard parts within each question.

Dealing with different types of question

Short-answer questions

These are the most straightforward, but remember:

- Look at the marks available — make one good point per mark.
- Look at the number of lines — this gives *some* idea of the length of answer required. Of course handwriting differs greatly in size, but if you have written two words and there are three lines, you may assume you have not written enough to score full marks.
- Don't 'hedge your bets' — if you give two alternative answers, you will not get the marks unless *both* are right. For example, if the answer is 'mass number' and you write 'mass number or relative atomic mass', you will score zero.
- Read the question — don't answer a question which you have made up. Examiners do have kind hearts really, and they are genuinely sorry when they have to award zero for an answer containing good chemistry that is not relevant to the question asked. This is a problem with units which are examined twice a year. There are lots of past papers around, all asking slightly different questions on the same subject-matter. It's all too easy to give the answer to last year's question.

Long-answer questions

The same rules apply about marks, lines and reading the question. In addition:

- Think before you write ('put brain into gear before operating hand') — perhaps jot a few points in the margin. Try to make your points logically.
- Punch those points — if you have read any mark-schemes you will see that they give the examiners advice on the weakest answer that will still just score the mark. Make sure your points are made well and win the mark without a second's hesitation by the examiner.
- Try to write in clear sentences (though bullet points might be better on some occasions).

- Be sure you do not re-state the question, i.e. don't use words or phrases directly from the question as part of your explanation.
- ✎ In your answer, you should use appropriate technical terms, spelt correctly.

There is 1 mark for quality of written communication in Unit F331, highlighted by the ✎ symbol and the words above. Examiners are usually relaxed about slight spelling errors, but in this question they can't be. Check your spelling carefully in this question.

Command words in questions

A lot of care is taken in choosing which of these words to use, so note them carefully:

- 'State', 'write down', 'give', 'name' require short answers only.
- 'Describe' requires an accurate account of the main points, but no explanation.
- 'Explain' requires chemical reasons for the statement given.
- 'Suggest' means that you are not expected to know the answer but you should be able to work it out from what you do know.
- 'Giving reason(s)' requires you to explain why answered as you did (if 'reasons' in the plural is stated, judge the number required from the number of marks).

Avoid vague answers

Sometimes it is clear that the candidate knows a lot about the topic but his or her answer is not focused. Avoid these words:

- 'It' (e.g. 'it is bigger') — give the name of the thing you are describing, otherwise it may not be very clear which object in the question is being referred to.
- 'Harmful' — if you mean 'toxic' or 'poisonous' say so.
- 'Environmentally friendly' — say *why* it benefits the environment.
- 'Expensive' — always justify this word with a reason.

Be careful with chemical particles — always think twice whenever you write 'particle', 'atom', 'molecule' or 'ion', and check that you are using the correct term.

If in doubt, write something

Try to avoid leaving gaps. Have a go at every answer. If you're not sure, write something that seems to be sensible chemistry. As you will see from the Question and Answer section, some questions have a variety of possible answers — the only answer that definitely scores zero is a blank.

Diagrams

You would be amazed at some of the diagrams examiners have to mark, so please:

- Read the question. The answer is not always a reflux condenser. If it is an apparatus you know, then it is relatively straightforward. If you have to design something, look for clues in the question.
- Make it clear and neat. Use a pencil and ruler, and have a soft rubber handy to erase any errors.
- Make sure it looks like real apparatus (which never has square corners, for example). Some apparatus drawn in exams would test the skill of the most proficient glass-blower.

- Draw a cross-section, so that gases can have a clear path through. Don't carelessly leave any gaps where gases could leak out.
- Think of safety. Don't suggest heating an enclosed apparatus, which will explode. If a poisonous gas is given off, show it being released in a fume cupboard.
- Always label your diagram, especially if the question tells you to. Important things to label are substances and calibrated vessels (e.g. syringes or measuring cylinders).

Calculations

If you get the answer to a calculation right, the working does not need to be there (unless you could have guessed the answer). However, this is the most misleading piece of information in this book. It is always very easy to make mistakes, especially under the pressure of exams. So, set out the steps in your calculations clearly. Then you will get most of the marks if you make a slight mistake and the examiner can see what you are doing. Examiners operate a system called 'error carried forward' whereby, once an error has been made, the rest of the calculation scores marks if the method is correct from then on.

At the end of the calculation, there will be a line that reads, for example:

Answer________ (2 marks)

Obviously you should write your answer clearly here. When you write down your numerical answer, check:

- **units** — most physical quantities have them. (Sometimes these appear on the answer line to help you)
- **sign** (remember that ΔH values must be shown as '+' if they are positive)
- **significant figures** — you may be expected to analyse uncertainties more carefully in your coursework, but in exam papers all you have to do is to give the same number of significant figures as the data in the question

On-screen marking

When you have written your answers, the papers are collected and sent to a scanning centre. There they are scanned so that they can be marked online by examiners. This hardly matters to you, but just be aware of the following requirements, which are signalled in the instructions on the front of the paper:

- Use *black* or *dark-blue* ink only.
- Use pencil only for diagrams and graphs. Make sure that the pencil is an HB type and makes a good black mark.
- Do not use colours, as these won't scan well and they will be scanned in various shades of grey anyway.
- Write clearly, as the scanning process makes it just a little more difficult to read 'dubious' writing (if you are unsure, try photocopying some of your work and see how clear this is).
- Try to keep your answer within the dots provided. If you have to cross out an answer or you run out of room, you may write it on the same page if there is space,

but keep your answer clear — don't write up the side of the page. If you run out of room, you may ask the invigilator for an extension sheet. Make it clear which question and part you are answering, of course, and write your name at the top, just in case the extension sheet becomes separated from the paper. Extension sheets will be scanned.

- Do not write on blank pages (where it says 'do not write on this page'), however tempting this may be. Blank pages may not be scanned.

However, be reassured that, if you make any mistakes with regard to any of these points, the examiners will do their best not to penalise you.

Content Guidance

The material in this section summarises the chemical ideas from **Elements of life** and **Developing fuels**. It is arranged in chemical order, rather than in the order in which you have studied it.

Summary of content

About the atom: the make-up of an atom, mass spectra of molecules, nuclear reactions and radioactivity, half-life, electron arrangement and atomic spectra, models of the atom.

The periodic table: history, periodic trends of physical and chemical properties, electron structure, similarities and trends in a group.

Bonding and structure: ionic bonding, covalent molecules, shapes of covalent molecules, metallic bonding, structure and properties.

Alkanes: the alkane series, isomerism, naming alkanes, other organic compounds, naming alcohols.

Petrol: characteristics of a good fuel, improving alkanes, octane rating, pollutants, which fuel for the future?

Catalysis: catalysis and cars, catalytic converters, zeolites.

Entropy: what is entropy?

Mole calculations and equations: moles, formulae, equations, state symbols, calculations from equations.

Enthalpy changes: definitions, measuring enthalpy changes, combustion of liquid fuels, reactions in solution, standard enthalpy changes, Hess cycles, bond enthalpies.

How much of this do I need to learn?

The answer is virtually all of it. It has been pared down to the absolute essentials. If you need any more detail on any aspect, you should look in your textbooks or notes.

About the atom

The make-up of an atom

The nucleus of an atom consists of **protons** and **neutrons**. Round the nucleus move the **electrons**. The evidence for this is discussed later (see p. 20). These particles are related as shown in the table below:

Particle	Mass	Charge
Proton	1	+1
Neutron	1	0
Electron	Very small	–1

The number of these particles in an atom is indicated like this:

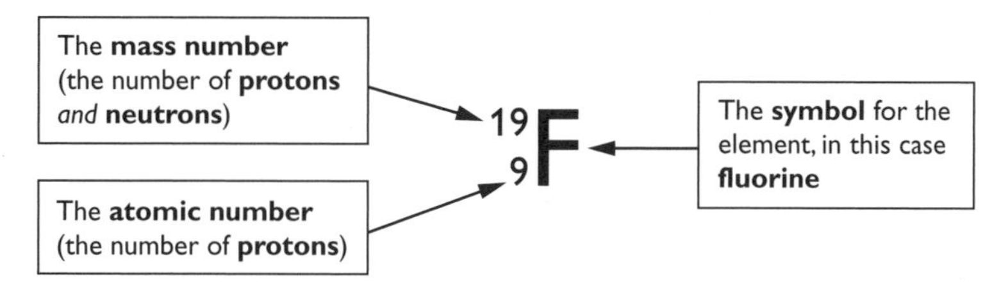

For a neutral atom, the number of electrons is equal to the number of protons.

Isotopes are forms of the same element. Their atoms have the *same number of protons* but *different numbers of neutrons*. For example, hydrogen has *three isotopes*, $^{1}_{1}H$, $^{2}_{1}H$ and $^{3}_{1}H$. These all have one proton (otherwise they wouldn't be hydrogen) but they have zero, one and two neutrons, respectively, as you can see from the mass numbers.

The **relative atomic mass** of an atom is defined as the number of times one atom is heavier than one-twelfth of an atom of carbon-12 (the isotope $^{12}_{6}C$). For an element that consists of just one isotope (for example, fluorine), this means that the relative atomic mass is virtually the same as the mass number: in this case 19.

However, when an element consists of two or more isotopes, the relative atomic mass is the mean (average) of the **relative isotopic masses** of the isotopes, taking into account their relative abundances (how much of each of them occurs naturally). For example, chlorine has two isotopes, $^{35}_{17}Cl$ (which makes up 75% of chlorine) and $^{37}_{17}Cl$ (which makes up the remaining 25%).

The relative atomic mass is calculated like this:

$$\frac{(35 \times 75) + (37 \times 25)}{100} = 35.5$$

Measuring atomic masses

The masses of atoms can be measured using an instrument called a **mass spectrometer**. This works as shown in the following diagram:

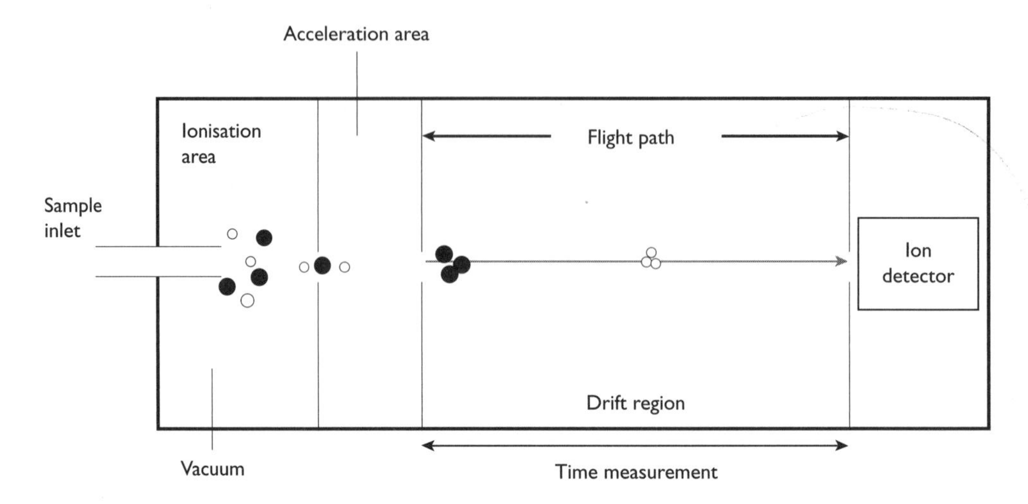

Tip You are more likely to be asked to label the diagram than to draw it.

Sample inlet: gaseous atoms of the sample enter.

Ionisation area: the atoms are bombarded with electrons that knock one electron off each atom, leaving 1+ ions.

Acceleration area: the ions are accelerated by an electric field. This is turned on for a very short time, in order to produce a pulse of ions all with the same kinetic energy.

Drift region: here the ions move under the same electrical field. The time that they take to cover the **flight path** is measured. Because all the ions have the same kinetic energy ($\frac{1}{2}mv^2$), the heavier ones will move more slowly. So the ions arrive at the ion detector in clumps, according to their mass.

The **time of flight** that is measured can be mathematically related to the mass of the ions (or, more exactly, m/z, where z is the charge on the ion — just in case some ions have acquired more than one charge in the ionisation area — for the AS course, z can always be assumed to be equal to 1, so m/z will be the same as 'm').

Detector: there are various means of detecting ions, and you don't need to know the detail. The ions produce a current, which measures their intensity.

The whole apparatus has to be under a very high **vacuum**, so that the ions do not collide with gas molecules. From the calculations on the times of flight, a **mass spectrum** is produced. The mass spectrum of bromine is shown on p. 15.

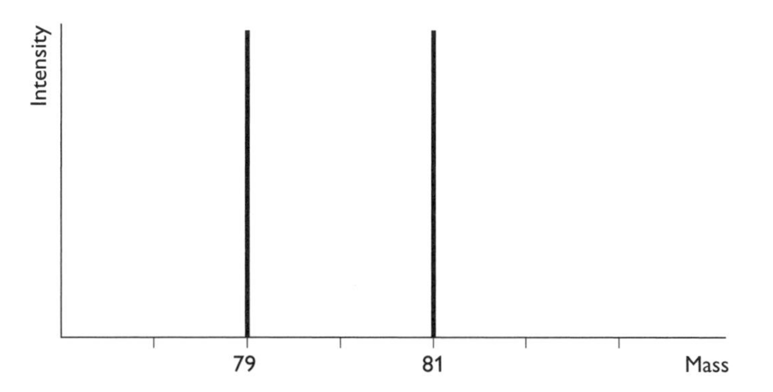

This shows that there are two isotopes, with masses of 79 and 81, of equal intensity. The relative atomic mass is given by:

$$\frac{(79 \times 50) + (81 \times 50)}{100} = 80$$

Mass spectra of molecules

Mass spectrometers are mainly used today to determine the structure of molecules. The molecules are ionised in the ionisation area. Some are broken into fragments. Some survive as molecules with one electron knocked off and are called **molecular ions**. Their mass gives the relative molecular mass of the whole molecule. The masses of the fragments can be used to work out the structure of the molecule, but you won't need to know the detail of this until you study A2 Unit F334.

The mass spectrum for ethanol, C_2H_5OH, is shown below:

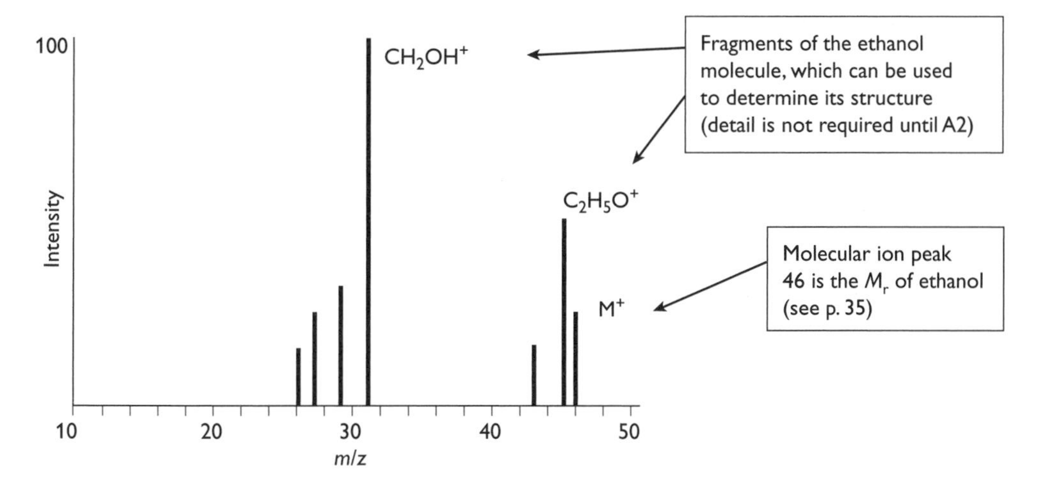

Nuclear reactions and radioactivity

The elements are made in stars like our Sun by **nuclear fusion**. This happens when lighter nuclei come together to make heavier ones. High temperature and pressure

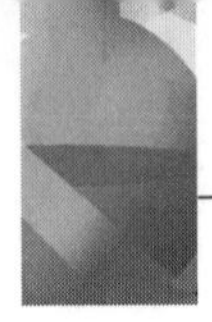

are needed to overcome the repulsion of the protons in the two nuclei that are fusing together. One example is:

$${}^{1}_{1}H + {}^{2}_{1}H \rightarrow {}^{3}_{2}He$$

where ${}^{1}_{1}H$ etc. are called **nuclear symbols**.

This is an example of a nuclear reaction. The rule is: *the sum of the lower numbers (atomic numbers) must be equal on each side of the equation, as must the sum of the upper numbers (the mass numbers)*. Any 'new' elements can be identified from the atomic numbers (e.g. element number two is always helium).

Tip In the examination you will always have the periodic table on your *Data Sheet* to consult.

Another reaction in which a heavier nucleus is formed is:

$${}^{9}_{4}Be + {}^{4}_{2}He \rightarrow {}^{12}_{6}C + {}^{1}_{0}n$$

Here a neutron is also formed — notice how it is represented in such a nuclear equation. (It is logical — no protons and one neutron.)

The protons and neutrons in some nuclei do not stay together. These nuclei are said to be **unstable** and **radioactive**. The three products of **radioactive decay** are shown in the following table.

Radiation[1]	What does it consist of?	Its penetrating power	Example of formation[2]
α-particle	Helium nucleus, ${}^{4}_{2}He$	Small — stopped by paper and human skin	${}^{235}_{92}U \rightarrow {}^{4}_{2}He + {}^{231}_{90}Th$ Radioactive decay of uranium-235
β-particle	Electron, ${}^{0}_{-1}e$ (note that the electron comes from the *nucleus* — it is not one of the electrons that surround the atom)	Medium — stopped by sheets of metal such as aluminium	${}^{14}_{6}C \rightarrow {}^{14}_{7}N + {}^{0}_{-1}e$ Radioactive decay of carbon-14
γ-ray	Electromagnetic radiation	High — nothing stops these rays, but their intensity is usually reduced to safe levels by several centimetres of lead	Formed during α or β decay

[1]All three are ionising radiations. They carry a lot of energy, so they can knock electrons off atoms in their path. This can cause damage to the genetic material in a cell, leading to cancer. Because they are ionising, they can be detected by a Geiger counter.

[2]The nuclear equations follow the same rule about the sums of numbers given above. Notice the symbol for an electron: ${}^{0}_{-1}e$

Tip You are not expected to memorise the examples in the final column of the table. You need to know how to represent the electron and the α-particle and how to balance nuclear equations.

Half-life

This is the time taken for half of the radioactive nuclei in a sample to decay. It is fixed for a given isotope and does not depend on the amount of isotope present. Half-lives vary from fractions of a second to millions of years.

Nitrogen-13 has a half-life of 10.0 minutes. This means that it decays as shown in the table below:

Time	Description	Amount remaining
10 mins	One half-life	One-half
20 mins	Two half-lives	One-quarter
30 mins	Three half-lives	One-eighth
40 mins	Four half-lives	One-sixteenth

(The graph on p. 18 illustrates this process for the decay of carbon-14.)

Radioactive tracers consist of radioactive atoms of an element that are used to follow the progress of the main bulk of the element, often around an animal body or plant structure. The position of the tracer can be found by the radioactivity that it is emitting. Isotopes emitting **β-particles**, with their medium penetrating power, are often chosen for use as tracers (β-particles have a low penetrating power and may be absorbed by the body). The half-life of the isotope used must not be too short, otherwise the radioactivity will not last for long enough. However, the radioactivity may be dangerous if it persists for a long period after the tracing has finished.

Tracers are used in dating archaeological and geological material. They often work on the principle that:

- while the sample is alive (or a rock is being formed from magma), it exchanges matter with the environment, so the amount of isotope remains constant
- when the sample dies (or the rock finishes forming), it ceases to exchange matter with the environment, so the isotope decays and is no longer replaced
- the date of the sample can be deduced from the remaining amount of the sample (measured by a mass spectrometer) and the known half-life
- in geological studies, the ratio of the radioactive isotope to a stable **daughter product** is used. Isotopes with half-lives of billions of years are used
- in archaeological studies, the ratio of an isotope (usually carbon-14) to a stable isotope (carbon-12) is used. Carbon-14 has a half-life of 5370 years

Example

The half-life of potassium-40 is 1.3×10^9 years. The stable daughter product of the decay of potassium-40 is argon-40. How old is a rock where the ratio of potassium-40 to argon-40 is 1:3?

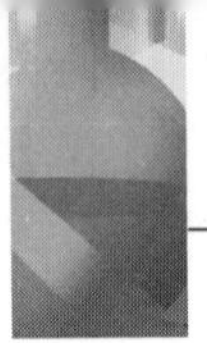

The ratio 1:3 means that 75% of the potassium-40 has decayed and 25% remains. This means that the rock is two half-lives old — that is 2.6×10^9 years.

Example

The half-life of carbon-14 is 5370 years. In living things, the ratio of carbon-14 to carbon-12 is 1×10^{-12}. Plot a graph of the ratio against time, and use it to date a piece of wood where the ratio is 0.60×10^{-12}.

The answer is as follows:

- The graph is plotted using the fact that each successive half-life is 5370 years (ratio 0.5×10^{-12} after the first half-life, 0.25×10^{-12} after the second, and so on). The solid lines show this.
- The points are then joined with a smooth curve.
- The time at a ratio of 0.6×10^{-12} is then read off (see the dashed lines). It comes to about 4000 years.

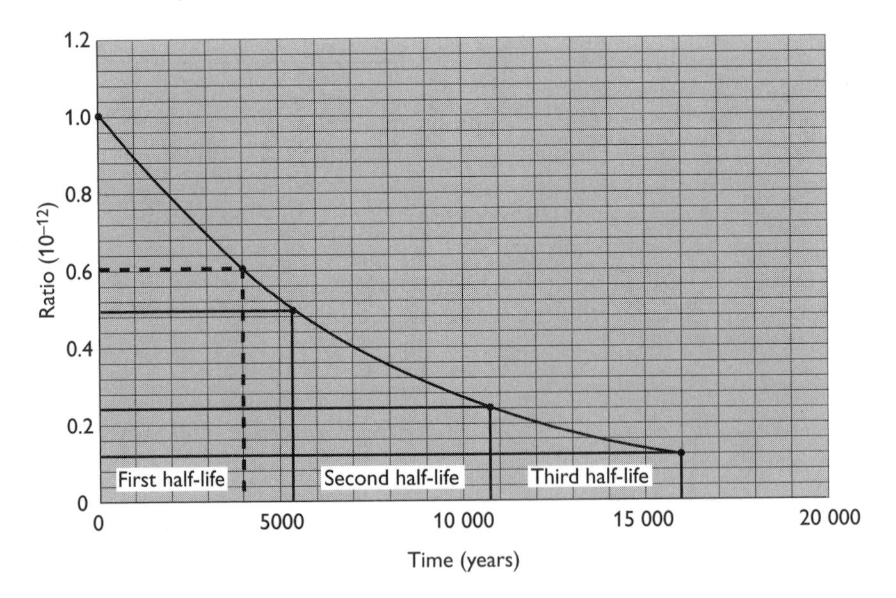

The following must be true if these methods of dating are to be used:

- The half-life of the radioisotope must be known accurately.
- There must be no movement of the radioisotope (or the daughter product) out of the material.
- In the case of rocks, there must have been no subsequent re-melting, which starts the clock all over again.

Electron arrangement and atomic spectra

The electrons in atoms are arranged in **shells** or **energy levels**. These determine the element's position in the periodic table, as shown below for the first 20 elements. The consequences of this are considered further in the sections on 'The periodic table' (p.21) and 'Bonding' (p.23).

H 1							He 2
Li 2.1	Be 2.2	B 2.3	C 2.4	N 2.5	O 2.6	F 2.7	Ne 2.8
Na 2.8.1	Mg 2.8.2	Al 2.8.3	Si 2.8.4	P 2.8.5	S 2.8.6	Cl 2.8.7	Ar 2.8.8
K 2.8.8.1	Ca 2.8.8.2						

Tip You will be expected to quote these arrangements, but you will always have the periodic table in your *Data Sheet* for guidance.

The fact that electrons can only exist in definite energy levels leads to the production of **atomic spectra**. The energy levels of a hydrogen atom are shown below — other atoms' energy levels are similar, though more complex.

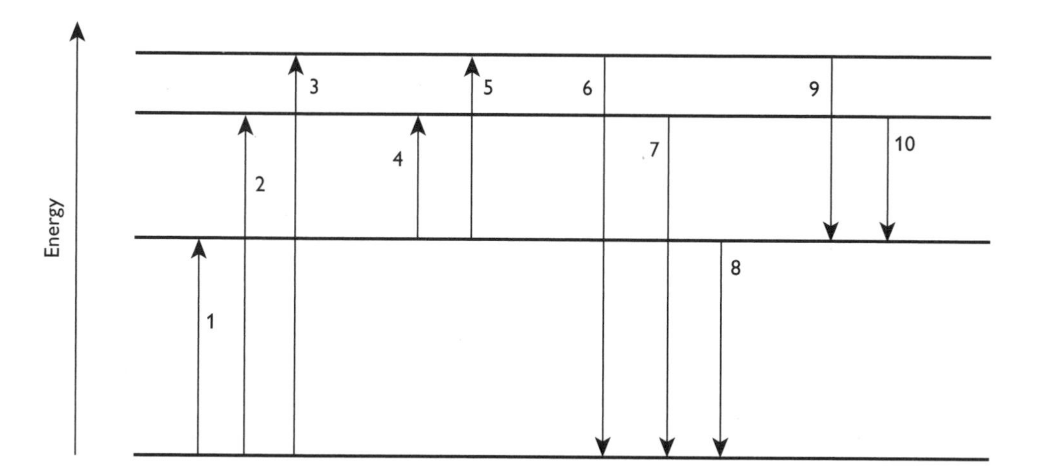

An electron in the lowest energy level (the **ground state**) can only go up to a higher energy level if it receives just the right amount of energy (1 in the diagram). Absorption of other specific amounts of energy moves the electron up further (2 and 3). If this energy comes from radiation such as light, it must have just the right frequency, with the energy being proportional to the frequency.

The expression $\Delta E = h\nu$ represents the relationship. ΔE is the separation between the energy levels, ν is the frequency, and h the Planck constant (more about this is given in Unit F332).

When electrons have been excited (moved up energy levels) by being heated, they then drop down to lower energy levels, giving out a frequency of light which is proportional to the energy difference through which they have fallen (e.g. lines 6 to 10). (The numbers of the lines correspond to the energy changes in the energy level diagram above.)

Series of converging lines all starting/finishing at the level above the lowest

Series of converging lines all starting/finishing at the lowest level

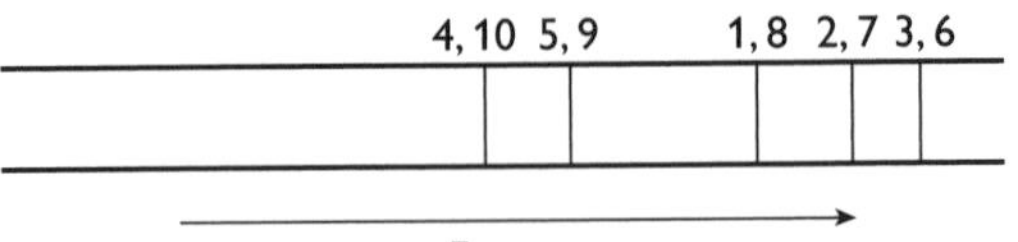

Frequency

Tip You may have to draw diagrams of energy levels and spectra like those above.

Absorption spectra

- are caused by electrons going up energy levels
- produce dark lines on a continuous ('rainbow-coloured') spectrum

Emission spectra

- are caused by electrons going down energy levels
- produce coloured lines on a dark background

Both

- are line spectra
- have the same lines in the same position for the same element
- have sets of lines corresponding to transitions to or from a particular level
- have each set of lines converging (getting closer at higher frequency) as the energy levels become closer together

Models of the atom

Tip You need to know these ideas in outline and be able to apply them to the development of the structure of the atom and other examples. You will always be given the necessary facts on which to comment.

- Scientists make models of complicated systems (like atoms) to help them to understand these systems better.
- As science progresses, these models become more sophisticated and can explain more of the behaviour of the system. However, they are still *models*. So we shouldn't ask 'What does an atom look like?', but rather 'What is the best model that we have of the atom?'
- Scientists test a model by doing experiments. If the results of their experiments can be explained by the model, they reinforce the model's usefulness. If their results cannot be explained by the model, the scientists suggest modifications to the model or, sometimes, a new model.
- These models (like all scientific knowledge) are *tentative*, that is they are suggestions based on what scientists know at the time.

The development of the structure of the atom by a series of more sophisticated models illustrates this. The table on p. 21 shows a few key models.

Date	Model	What it explained
1807	Dalton proposed that atoms were solid spheres	Formulae as simple ratios
1897	Thompson proposed that atoms contained electrons (arranged 'like currants in a plum pudding')	'Cathode rays' from low-pressure gases
1911	Rutherford postulated that the atom has a small, dense nucleus surrounded by electrons	Geiger and Marsden's experiment: α-particles were fired at gold foil. Most went through undeflected (these hadn't hit a nucleus); others 'bounced back' (these had hit a dense nucleus)
1914	Bohr proposed the idea of electron shells	Atomic spectra

Tip You do not need to learn these models, but you should understand how the ideas developed.

The periodic table

The periodic table lists the elements in order of **atomic number**, and groups elements together according to their common properties.

History

In the late seventeenth century, chemists were trying to classify the elements. The story is another one of a model which was gradually improved. There were several attempts to organise the elements based on relative atomic mass, but early versions did not fit the experimental facts well. The most successful arrangement was that of Mendeleev in 1871. His table was similar to that used today. However, he had to leave gaps for certain elements, which he said were not yet discovered, and he even predicted their properties. People ridiculed him for leaving these gaps. He also had to reverse the order of some elements (e.g. Te, I) so that they fitted the pattern of properties. Again, some people ridiculed him for this. However, some scientists set out to validate Mendeleev's table. The discovery of Ga and Ge, which fitted two of his gaps and had properties similar to those predicted, began his vindication. Later, when atomic number (rather than atomic mass) was used to order the elements, this placed Te and I in Mendeleev's order naturally, so he was shown to be right here too.

Tip Scientific research is used in this way to validate many discoveries. You may be asked to comment on another example. However, you will always be given the full facts.

Periodic trends of physical and chemical properties

Melting points and **boiling points** of the elements are said to show **periodic trends**, since the pattern of rise and fall is repeated as you go across the elements

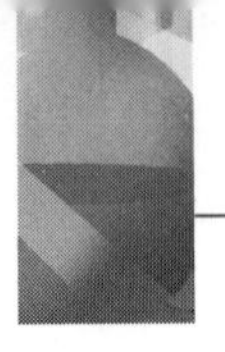

of each period. This is because the bonding of the elements changes from **metallic** to **giant covalent** to **covalent molecules** on moving from left to right across a period (see the section on 'Bonding').

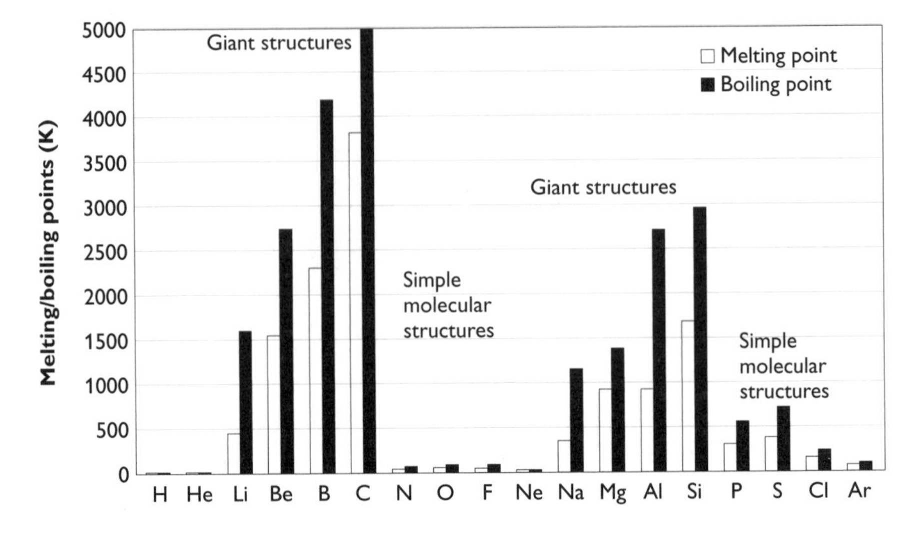

Tip You only need to know the outline *shapes* of these charts, *not* the values.

Group 2

As we go down the group from magnesium to calcium, strontium and barium, the elements become more metallic and their reactivity as metals increases. The properties you need to know about are:

- reactions with water
- acid–base character of the oxides and hydroxides
- thermal stability of the carbonates
- solubilities of hydroxides and carbonates

Reactions with water

The elements get *more reactive* down the group:

$$M(s) + 2H_2O(l) \rightarrow M(OH)_2(aq) + H_2(g)$$

Oxides and hydroxides

The oxides all react with water to form *alkaline* hydroxides:

$$MO(s) + H_2O(l) \rightarrow M(OH)_2(aq)$$

The oxides and hydroxides are also *bases*, since they react with acids to neutralise them. For example:

$$MO(s) + 2HCl(aq) \rightarrow MCl_2(aq) + H_2O(l)$$

and

$$M(OH)_2(s) + 2HCl(aq) \rightarrow MCl_2(aq) + 2H_2O(l)$$

The hydroxides become *more soluble* down the group.

Carbonates

All the group 2 carbonates decompose on heating:

$$MCO_3(s) \rightarrow MO(s) + CO_2$$

Barium carbonate needs the highest temperature, so it is said to have the highest **thermal stability**.

All the carbonates are insoluble, but they become *less soluble* down the group (note that this the opposite way round to the hydroxides).

Bonding and structure

Ionic bonding

When a metal and a non-metal combine, they form an **ionic compound**. Elements combine in order to lose energy. Often the ions formed have the full electron arrangement of the nearest noble gas. Metals (being on the left of the periodic table) can most easily do this by *losing electrons* to form positive **cations**. Non-metals *gain electrons* to form negative **anions**. The charges on common ions are shown in the table below and are well worth learning (you will need them for Unit F332).

The charges on ions formed by a single element can be related to their position in the periodic table. Carbonate, nitrate, sulfate and ammonium contain, of course, more than one non-metal, and their formulae have to be learnt — no chemist should be ignorant of these.

H **H^+** and **H^-**							He
Li **Li^+**	Be	B	C	N **N^{3-}**	O **O^{2-}**	F **F^-**	Ne
Na **Na^+**	Mg **Mg^{2+}**	Al **Al^{3+}**	Si	P **P^{3-}**	S **S^{2-}**	Cl **Cl^-**	Ar
K **K^+**	Ca **Ca^{2+}**		**CO_3^{2-}** carbonate	**NO_3^-** nitrate **NH_4^+** ammonium	**SO_4^{2-}** sulfate		

When forming ionic compounds, atoms exchange electrons so that both gain a full outer shell of electrons. This can be shown by a **dot-and-cross diagram**, which shows just the outer shell of electrons.

For example, magnesium has two outer-shell electrons. It can give one electron (shown as × below) to each of two chlorine atoms to make magnesium chloride, $MgCl_2$.

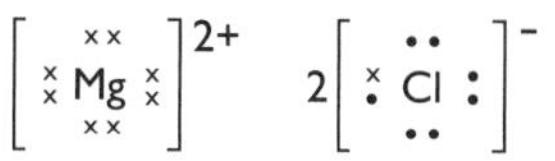

$$[\text{Mg}]^{2+} \quad 2[\text{Cl}]^{-}$$

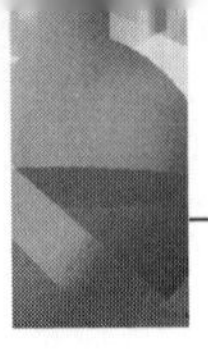

Covalent molecules

When two non-metals combine they *share* electrons to form a **covalent** bond. Dot-and-cross diagrams can be used to illustrate this too. The vital things to remember are:

- the number of outer-shell electrons in an atom is the same as its group number
- the number of electrons in the outer shell of second-period atoms (Li to Ne) is limited to eight
- a complete outer shell of eight electrons is a very stable arrangement (hence the unreactivity of neon)
- elements in later periods can 'expand their octets' (that is, have up to 18 electrons around them)

Atoms in the second period share electrons to gain these outer shells of eight (or two for hydrogen). Any non-bonding electrons which are not needed are shown as paired up (and are known as **lone pairs**). Examples are shown below.

Tip You should be able to draw these and other examples.

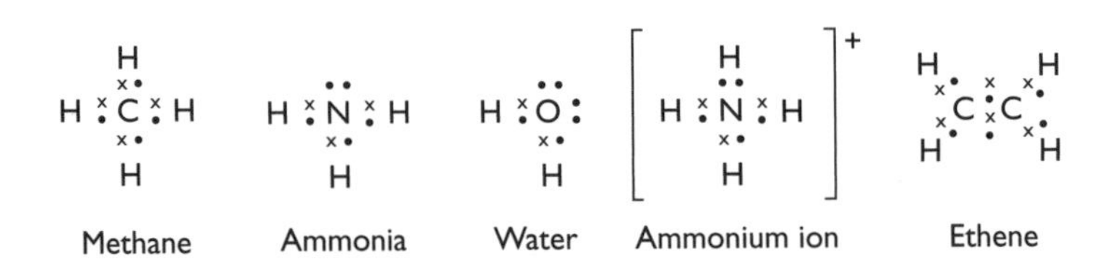

Note that in the ammonium ion, both electrons for one of the bonds (to the H^+ ion) come from the nitrogen. This is called a **dative covalent bond** (because both the electrons are *given* by the nitrogen atom). Ethene is a simple example of a molecule containing a **double bond**, which is shown as two shared pairs.

Dot-and-cross diagrams are *models*. They only help us to account for the numbers of electrons. They do not:

- indicate how the electrons are arranged in space
- indicate the shape of molecules or ions

The theory can be extended to include shape (as below) but, to explain more advanced applications of bonding, more sophisticated models are needed. *However, you do not need to know about them.*

Shapes of covalent molecules

From the dot-and-cross diagrams, the shapes of molecules can be deduced by remembering the simple rules:

- Lone pairs, bonding pairs and double bonds all count as 'groups of electrons'.
- Groups of electrons repel each other and get as far away from each other as they can.

- When there are *four* groups of electrons, the **bond angle** is approximately 109° (the angle of a regular **tetrahedron**). For example:

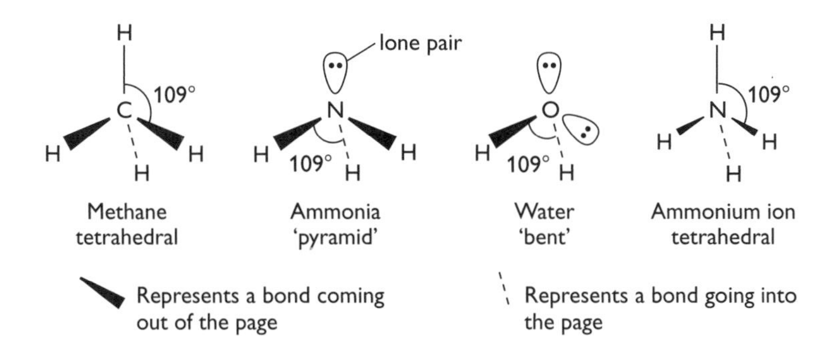

- When there are three groups of electrons, the structure is described as trigonal (triangular) and the bond angle is approximately 120°. For example:

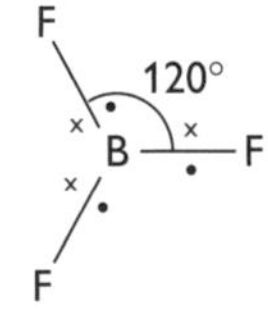

Boron trifluoride

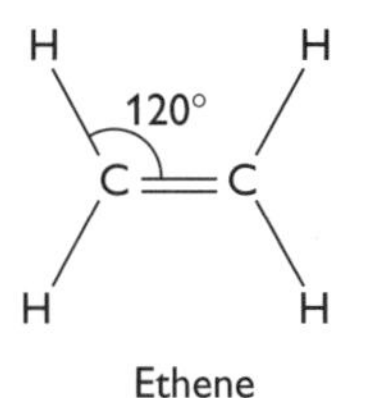

Ethene

Note: boron has only three electrons to pair up

Note: the double bond counts as *one* group of electrons

- When there are *two* groups of electrons, the structure is described as **linear** and the bond angle is 180°. For example:

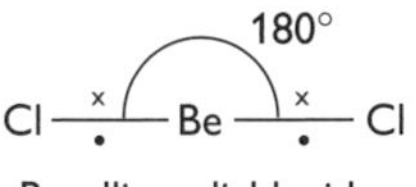

Beryllium dichloride

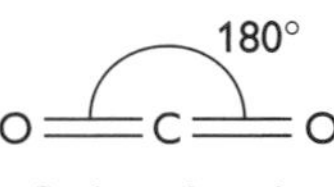

Carbon dioxide

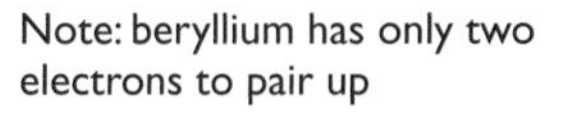
Note: beryllium has only two electrons to pair up

- When there are *six* groups of electrons, the structure is described as **octahedral**. An octahedron has eight *faces* (but only six points). The bond angle is 90°. Sulfur forms a fluoride, SF_6, that shows this symmetry:

F
F S F
F F
F

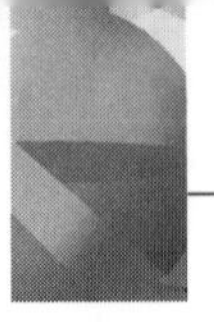

Sulfur has six electrons in its outer shell. It can share each of these with an electron from fluorine, making SF_6. (Because it is in the third period, it can expand its octet and have more than eight electrons — up to 18 — in its outer shell.)

The shape of SF_6 is shown below:

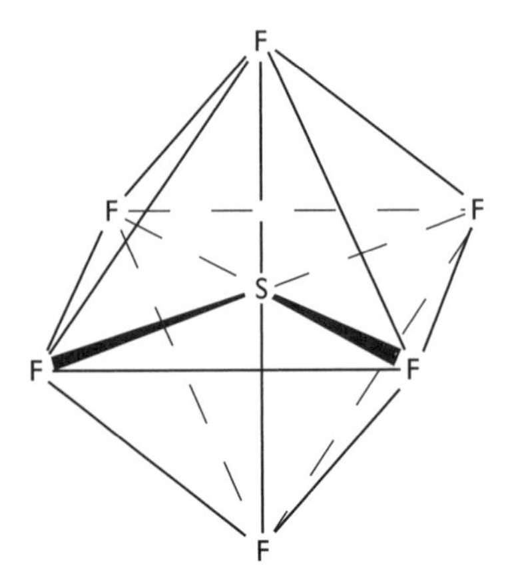

(Note that in this diagram the three lone pairs on each F atom are omitted for clarity.)

Metallic bonding

In the metal structure, the outer-shell electrons of metal atoms come free from the atoms. Thus, metals consist of a **lattice** of positive ions. This accounts for their ability to bend and stretch without breaking, since the ions can move over each other in layers.

The electrons can move around between the ions and are said to be **delocalised**. This causes the metal to conduct electricity. Think of 'cation islands in a sea of electrons'. The positively charged ions attract the negative electrons, and this attraction holds the structure together.

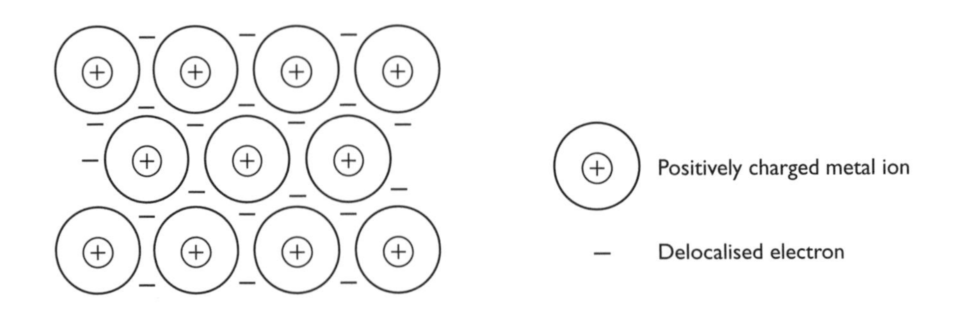

Structure and properties

Structure type	Bonding	Example	Melting point	Solubility in water	Ability to conduct electricity
Giant lattice					
Metallic	Metallic	Magnesium	High	Insoluble	Conduct when solid
Ionic	Ionic	Sodium chloride	High	Often soluble	Conduct only when molten or in solution
Covalent network	Covalent	Diamond	Very high	Insoluble	Non-conductor
Simple molecular	Covalent	Methane	Low	Often not very soluble	Non-conductor

Tip You need to know these facts, but not (yet) the reasons.

Alkanes

Alkanes are the series of hydrocarbons with general formula $C_nH_{(2n+2)}$. This is an example of a homologous series. The first member of the series is methane. Three ways of representing the formula of methane are shown below:

Molecular formula

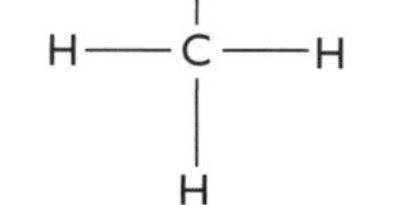

Full structural formula

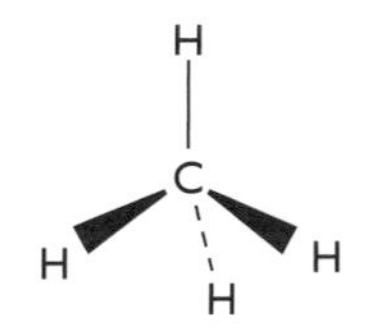

Three-dimensional structural formula

Then come:

Ethane	C_2H_6	Heptane	C_7H_{16}
Propane	C_3H_8	Octane	C_8H_{18}
Butane	C_4H_{10}	Nonane	C_9H_{20}
Pentane	C_5H_{12}	Decane	$C_{10}H_{22}$
Hexane	C_6H_{14}	Undecane	$C_{11}H_{24}$

Tip You should learn the names of the alkanes up to decane.

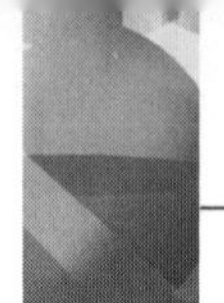

Isomerism

For butane and above, structural isomerism is possible. **Structural isomers** have the same molecular formula but a different structural formula. The shortened structural, full structural and skeletal formulae of the isomers of butane and pentane are shown in the table below.

Name	Shortened structural formula	Full structural formula	Skeletal formula
Isomers of C_4H_{10}			
Butane	$CH_3-CH_2-CH_2-CH_3$		
Methylpropane	CH_3 | $CH_3-CH-CH_3$		
Isomers of C_5H_{12}			
Pentane	$CH_3-CH_2-CH_2-CH_2-CH_3$		
2-methylbutane	CH_3 | $CH_3-CH-CH_2-CH_3$		
2,2-dimethylpropane	CH_3 | CH_3-C-CH_3 | CH_3		

Notes on structural formulae

- **Shortened structural formulae** can be written in a variety of ways. They show the structure but they do not display all the bonds.
- **Full structural formulae** show *all* the bonds and atoms. Remember this if you are drawing one.
- **Skeletal formulae** look strange at first but they are used by chemists to represent larger structures. At the end of each line and at each angle, there is assumed to be a carbon atom with the right number of hydrogens attached.

Naming alkanes

The rules are:

- Identify the longest carbon chain. (This can sometimes 'go round corners'.) The number of atoms in this chain gives the basic name — propane, butane, pentane etc.
- Use the names methyl (CH_3), ethyl (C_2H_5) etc. to describe the side-branches.
- Name the compound, giving the number of the carbon atom on which each side-branch hangs, counting from whichever end of the chain gives the smaller number. For example, 2-methylpentane means that a five-membered chain has a methyl group on the second atom along.

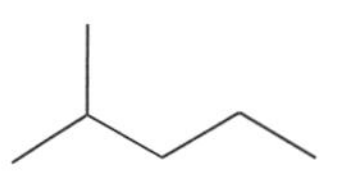

- When there are two branches on the same carbon atom, it is written like this: 2,2-dimethylpropane.

Tip Note that numbers are separated by commas; numbers and words are separated by hyphens; and there is no gap between words.

Thus, the structure shown below is called 2,2,4-trimethylpentane.

Other organic compounds

As well as alkanes, you are expected to recognise the homologous series listed in the table on p. 30.

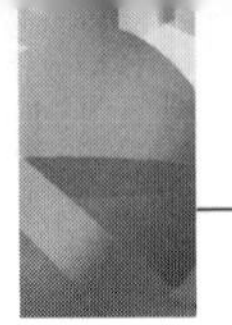

Name	Shortened structural formula	Full structural formula	Skeletal formula[1]
Cycloalkanes (e.g. cyclohexane — rings of carbon atoms)	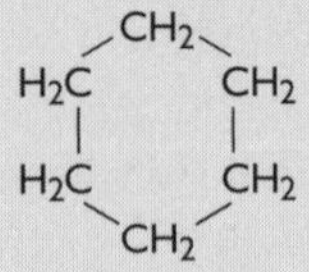	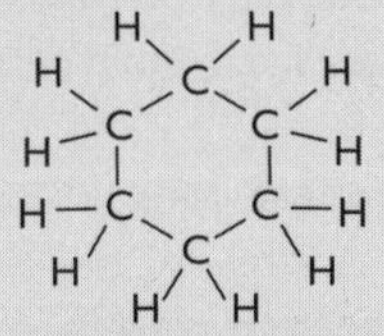	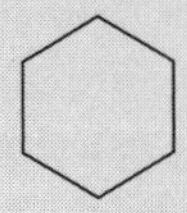
Alkenes (e.g. propene — one double bond)	$CH_2{=}CH{-}CH_3$	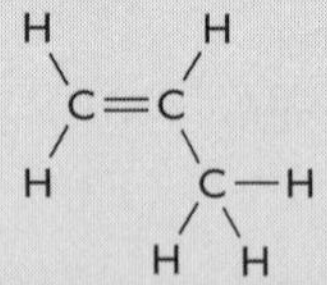	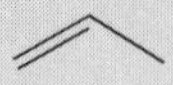
Alcohols (e.g. propan-1-ol — contain an –OH group (see below))	$CH_3{-}CH_2{-}CH_2{-}OH$	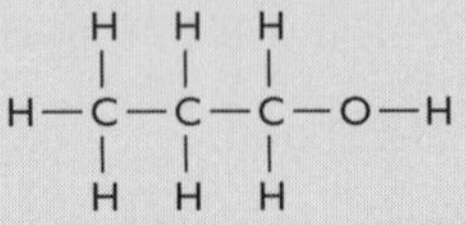	 OH
Ethers (e.g. diethyl ether — contain a C–O–C link	$CH_3{-}CH_2{-}O{-}CH_2{-}CH_3$	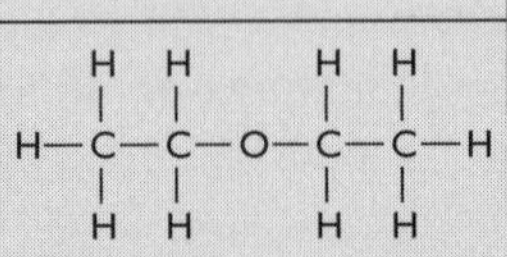	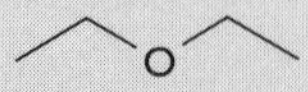 O

[1]In skeletal formulae, the bond to oxygen is shown and hydrogen atoms attached to the oxygen atoms are also shown. This is also the case for skeletal formulae when there are nitrogen atoms in organic compounds.

Tip You are not expected to be able to name compounds apart from alkanes and alcohols (see below).

The compounds in the table above are all called **aliphatic** to distinguish them from compounds containing **benzene rings**, which are called **arenes** or **aromatic compounds**. Benzene has the molecular formula C_6H_6 and is represented by the following symbol:

Naming alcohols

The rules for naming alcohols are:

- Find the longest carbon chain (as for alkanes). The number of carbon atoms gives the first part of the name: ethan-, propan-, butan- etc.
- The position of the OH group is described by the number of the carbon atom to which it is bonded, for example, propan-2-ol, 3-methylbutan-1-ol etc.

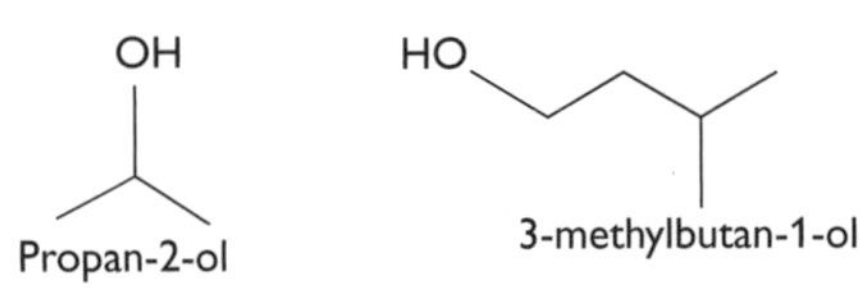

Petrol

Characteristics of a good fuel

Petrol consists of alkanes and other molecules. A good fuel:

- has the correct **volatility** to work in the engine
- has a high **energy density** ($kJ\,kg^{-1}$), i.e. it gives out a lot of energy per kilogram when it burns in air
- burns easily but does not 'auto-ignite' in the engine, i.e. it has the correct **octane number**
- does not corrode the engine
- produces the minimum level of polluting gases

Improving alkanes

Alkanes come from the fractional distillation of crude oil. This 'straight-run gasoline' can be made to perform better in engines by isomerisation, reforming and cracking.

Isomerisation means forming an isomer (usually more branched), for example:

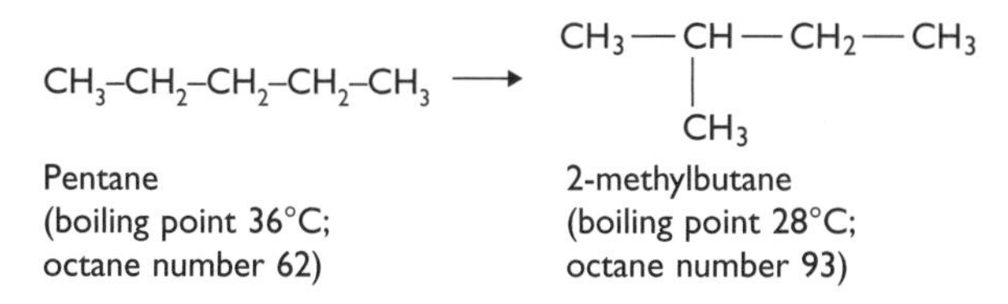

Reforming involves forming a ring compound from a chain (note that hydrogen is also produced), for example:

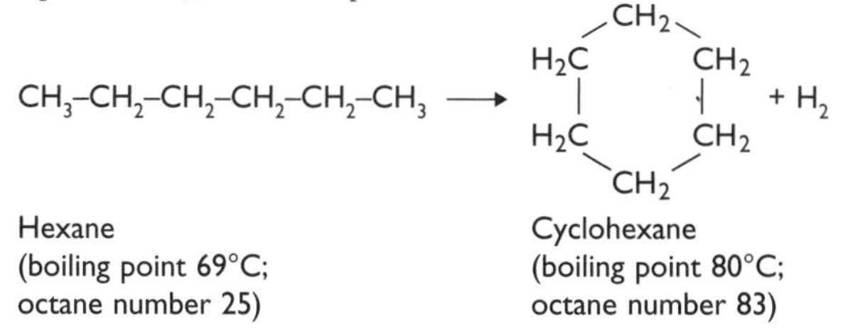

Cracking means breaking a long-chain alkane up into shorter chains, some with double bonds and some branched, for example:

CH_3–CH_2–CH_2–CH_2–CH_2–CH_2–CH_2–CH_2–CH_2–CH_2–CH_2–CH_3
Dodecane (boiling point 196°C — too high for petrol)

↓

CH_3—CH(CH_3)—CH_2—CH_3 + CH_3—C(CH_3)$_2$—CH_2—CH=CH_2

2-methylbutane
(boiling point 28°C; octane number 93)

4,4-dimethylpent-1-ene
(boiling point 63°C; octane number 144)

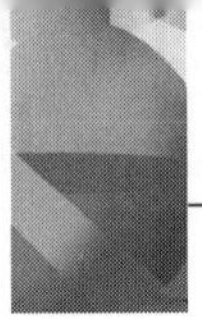

All these processes require **catalysts** (see p. 33). They produce compounds of a more suitable **volatility** and **octane number**.

Octane number

The **octane number** or **octane rating** of a petrol is a measure of how resistant it is to **pre-ignition** in a vehicle engine. In a modern, high-compression engine, the petrol is highly compressed in the cylinder before being sparked. Fuels of a low octane number explode without being sparked (**auto-ignition**) and cause a juddering or knocking in the engine. The octane number of a fuel can be improved by:

- decreasing the chain length
- increasing the amount of chain branching
- making a ring compound
- making an unsaturated (double-bond) compound
- adding **oxygenates** (alcohols or ethers) which have high octane numbers.

Pollutants from vehicles

Pollutant	Main effect	How is it produced?	How is it reduced?
Carbon monoxide, CO	• Toxic (stops blood taking up oxygen)	• Incomplete combustion of hydrocarbon fuels	• Adjust fuel/oxygen ratio • Use catalytic converters • Adding oxygenates is also thought to help
Particulates (microscopic carbon particles)	• Harm the lungs	• Incomplete combustion of diesel fuel	• Use diesel particulate filter
Nitrogen monoxide, NO (or NO_x)	• Toxic • Gives rise to acid rain which damages plants and buildings	• Nitrogen and oxygen from the air react in the heat of the engine	• Use catalytic converters
Unburnt hydrocarbons	• Combine with other gases to produce photochemical smog, which is a health hazard (especially for asthmatics)	• Emission in exhaust gases (incomplete combustion of fuel) and (by evaporation) from the engine and petrol tank • Also released while filling the car	• Adjust the fuel/oxygen ratio. • Better engine design • Use catalytic converters • Better petrol tank and petrol pump design
Sulfur dioxide, SO_2 (or SO_x)	• Gives rise to acid rain	• Burning of sulfur compounds in fuel	• Remove sulfur compounds from fuel (most are removed nowadays, so less of a problem)
Carbon dioxide, CO_2	• Greenhouse gas	• Burning of all hydrocarbon fuels	• Use more energy-efficient fuels • Use alternative non-hydrocarbon fuels

Tip You need to know all the details in this table.

Which fuel for the future?

Society has to make choices based on a whole range of advantages and disadvantages associated with the various fuels that are available. A brief summary of these is given below.

Method of energy production	Advantages	Disadvantages
Burning fossil fuels	• Can be used directly to power vehicles • Still cheap compared with most of the others	• Particulates, CO, CO_2 and sulfur compounds in the exhaust • Sources are not sustainable (they are running out)
Burning hydrogen	• Can be used directly to power vehicles • No particulates, CO, CO_2 or sulfur compounds in the exhaust	• NO_x still produced • Must be made using another energy source (for example, by electrolysis)
Burning biofuels – for example, ethanol made from fermenting starch; 'biodiesel' made by chemically treating plant oils	• Can be used directly to power vehicles • No particulates, CO or sulfur compounds in the exhaust • Renewable (can be grown), thus are sustainable	• NO_x still produced • CO_2 produced (though CO_2 has been absorbed during thr growth of the plants) • Uses up land needed for food production
Nuclear power	• No particulates, CO, CO, NO_x or sulfur compounds	• Cannot be used directly to power most vehicles • Decommissioning of power stations is expensive • Some danger due to leakage, waste etc.
Wind, wave, solar, hydro	• No particulates, CO, CO_2, NO_x or sulfur compounds • Sustainable (energy sources will not run out)	• Cannot be used directly to power vehicles • Large areas/high construction costs needed for a relatively small amount of energy (hydro is most effective, although suitable sites are relatively rare)

Tip You need to know the arguments here.

Catalysis

Catalysts and cars

A **catalyst** speeds up a reaction but can be recovered chemically unchanged at the end. Catalysts are used in two ways in connection with cars:

- **Catalytic converters**, usually made of powdered platinum and rhodium on a ceramic support, are used to catalyse the oxidation of carbon monoxide and

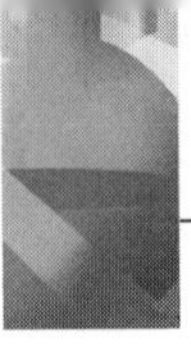

unburnt hydrocarbons to carbon dioxide (and water). They also catalyse this important reaction, which removes both carbon monoxide and oxides of nitrogen:

$$2NO + 2CO \rightarrow N_2 + 2CO_2$$

- Catalysts are used to speed up cracking, isomerisation and reforming reactions.

In both cases, the reactants and products of the reactions are *gases*, whereas the catalyst is a *solid*. This type of catalysis (where the catalyst is in a *different state* from the reactants) is called **heterogeneous catalysis**. The mechanism of action of such a catalyst is often as follows:

- Reactants are **adsorbed** (chemically bound) on to the catalyst surface.
- Bonds in the reactants weaken and break.
- New bonds are formed between atoms from the reactant molecules.
- The products diffuse away from the catalyst surface.

A catalyst is said to be **poisoned** if another substance is more strongly adsorbed on to the catalyst surface and stops it functioning. Lead compounds poison catalytic converters, so lead-free petrol had to be widely available before catalytic converters could be developed.

Zeolites are special catalysts that allow only molecules of special shapes to get through them. For example, they can separate branched and non-branched alkanes in **isomerisation** reactions.

Entropy

What is entropy?

Entropy is a measure of **disorder**. The more ways a set of particles can be arranged, the higher is its entropy.

A solid has least entropy, as there are relatively few ways of arranging its particles (high order). As we go from solid to liquid to gas, the entropy *increases*. Solutions often have high entropy too, as there are usually many different ways of arranging the particles.

Entropy increases in reactions in which more molecules of gas are produced.

A *mixture* of two substances has a higher entropy than the two substances separately. Again, there are more ways of arranging the particles of the substances when they are mixed.

Tip Try to name the particles concerned if describing entropy, but never confuse molecules and ions.

Mole calculations and equations

Moles

A **mole** is a way of counting, like a dozen or a score. A dozen atoms is 12 atoms. A mole of atoms is also a number: 600 000 000 000 000 000 000 000 atoms (much better written as 6×10^{23}). This large number is called the **Avogadro constant** (the number of particles in a mole). It is represented as N_A.

Remember that the **relative atomic mass (A_r)** of an element is the number of times one atom of that element is heavier than one-twelfth of an atom of carbon-12. The A_r for magnesium is 24. Thus, if there were 12 atoms of carbon in one beaker and 12 atoms of magnesium in another, the ratio of masses would be 1:2. If enough carbon atoms were added to one beaker to make 12 g and the *same number* of magnesium atoms were added to the other beaker, they would still be in the mass ratio 1:2 and both would contain the same number of atoms. There would now be one mole of each.

The mass of one mole of an atom is the A_r expressed in grams. So, for atoms:

$$\textbf{moles of atoms} = \frac{\textbf{mass (g)}}{A_r}$$

The **relative formula mass (M_r)** of a compound is the sum of the *relative atomic masses making up the formula*. For example: ethanol, C_2H_5OH, has:

$2 \times$ C atoms = $2 \times 12 = 24$
$6 \times$ H atoms = $6 \times 1 = 6$
$1 \times$ O atom = $1 \times 16 = 16$
Total = 46

So we say $M_r = 46$. This means that 46 g of ethanol contains the **Avogadro constant** of formula units, C_2H_5OH. (Since ethanol is made up of molecules, 46 can also be called the **relative molecular mass**, but **relative formula mass** is more general and applies to all formulae.)

So, for compounds:

$$\textbf{moles of formula units} = \frac{\textbf{mass (g)}}{M_r}$$

Formulae

Formulae can be worked out from combining masses by calculating the ratio of moles. Since all moles contain the same number of atoms, this gives the ratio of atoms.

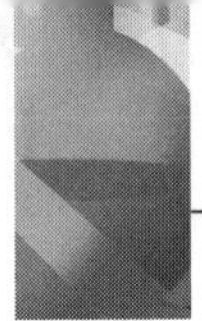

Example 1

A hydrocarbon contains 3.0 g of carbon and 1.0 g of hydrogen. Calculate its formula. The formula can be worked out like this:

	C	H
Mass	3.0 g	1.0 g
Moles	3.0/12.0 = 0.25	1.0/1.0 = 1.0
Ratio (divide by smaller)	1	4

Thus the formula is CH_4. This is called the **empirical formula**, since it is the simplest ratio of the atoms. In this case, it is also the **molecular formula** of methane, that is, the actual formula of the molecule. You should always look up the A_r values on the periodic table on the *Data Sheet*. Work to one decimal place and note that, while many A_r values are whole numbers to this accuracy, some are not — for example, sulfur is 32.1.

Example 2

Calculate the formula of a hydrocarbon containing 85.7% carbon. This can be done in a similar way, provided it is realised that the remaining 14.3% must be hydrogen. The percentages mean that 85.7 g of carbon combine with 14.3 g of hydrogen.

	C	H
Mass	85.7 g	14.3 g
Moles	85.7/12.0 = 7.14	14.3/1.0 = 14.3
Ratio (divide by smaller)	1	2

Here the empirical formula is CH_2. Bonding theory tells us that this is not possible, so we try C_2H_4. This is, of course, possible, as are C_3H_6, C_4H_8 etc., so we see that CH_2 is the empirical formula of all members of the alkene homologous series.

Example 3

What is the percentage of nitrogen by mass in the fertiliser ammonium nitrate (NH_4NO_3)? This is almost the reverse of the previous method. First, work out the M_r of NH_4NO_3:

$$14 + 4 + 14 + 48 = 80$$

There are two nitrogen atoms (= 28). Thus:

$$\text{percentage of nitrogen} = \frac{28 \times 100}{80} = 35\%$$

Equations

Equations show that atoms are just re-arranged in chemical reactions, not created or destroyed. So, equations must *balance* — there must be the same number of each atom on each side.

For example, methane burns in oxygen to form carbon dioxide and water. If you write down the formulae of the reactants and products as:

$$CH_4 + O_2 \rightarrow CO_2 + H_2O$$

you can see that the equation is *unbalanced.*

We cannot change the formulae, but we can change the number of moles reacting, that is, the numbers in front of the individual molecules' formulae.

Putting '2' in front of H_2O will give four hydrogens on each side. The carbon atoms are already balanced, so now it is just the oxygens that are unbalanced. Having doubled the water molecules, there are now four oxygen atoms on the right, so we need $2O_2$ on the left to balance:

$$CH_4 + 2O_2 \rightarrow CO_2 + 2H_2O$$

Sometimes, an odd number of oxygen atoms is needed in these circumstances. Then it is permissible to write the equation like this:

$$2C_3H_7OH + 9O_2 \rightarrow 6CO_2 + 8H_2O$$

or

$$C_3H_7OH + 4.5O_2 \rightarrow 3CO_2 + 4H_2O$$

Tip It *is* permissible to use halves (more so at AS than GCSE). Note that when balancing the equation for the combustion of an alcohol, there is one oxygen atom in the alcohol formula. This is very easy to miss.

State symbols

The **state symbols** are:

- (g) gas
- (l) liquid
- (s) solid
- (aq) aqueous solution

State symbols add more information to an equation:

$$Mg(s) + 2HCl(aq) \rightarrow MgCl_2(aq) + H_2(g)$$

Tip You need only add state symbols to an equation when the question specifically asks for them. Otherwise it is safest to leave them out.

Calculations from equations

The numbers in front of formulae in equations indicate how many moles are reacting. So the rule is: *turn mass of the given substance into moles, use the ratio from the equation to calculate moles of the required substance, then turn the moles back to masses.*

Example 4

Calculate the mass of sodium carbonate which is formed when 16.8 g of sodium hydrogen carbonate is heated. (A_r: Na, 23; C, 12; H, 1.0; O, 16, from the *Data Sheet*)

The equation is:

$$2NaHCO_3 \rightarrow Na_2CO_3 + H_2O + CO_2$$

The M_r of $NaHCO_3$ is (23 + 1 + 12 + 48) = 84. So moles of $NaHCO_3$ = 16.8/84.0 = 0.20.

The equation shows that two moles of $NaHCO_3$ form one mole of Na_2CO_3. So 0.20 moles forms 0.10 moles of Na_2CO_3. This has a mass of 10.6 g (M_r of Na_2CO_3 = 106).

A slight variation on this uses the fact that *one mole of molecules of a gas occupies 24 dm³ at room temperature and pressure* (you will be given this information when you need it).

So, in Example 4, if we needed to calculate the volume of carbon dioxide formed, we could have noted that the same number of moles of CO_2 are formed as Na_2CO_3, in this case 0.10. Thus, $0.10 \times 24 = 2.4\ dm^3$ of gas would be formed at room temperature and pressure.

If we are dealing with reactants and products that are *all gases*, the volumes are proportional to the moles reacting. Thus, for methane burning:

$$CH_4(g) + 2O_2(g) \rightarrow CO_2(g) + 2H_2O(g)$$

Here, one volume of methane (say $1.0\ dm^3$) would react with two volumes of oxygen ($2.0\ dm^3$) to form one volume ($1.0\ dm^3$) of carbon dioxide.

Enthalpy changes

Definitions

When chemical reactions occur, heat may be given out or absorbed. The stored chemical energy which accounts for this is called **enthalpy**. **Enthalpy level diagrams**, such as those below, illustrate this.

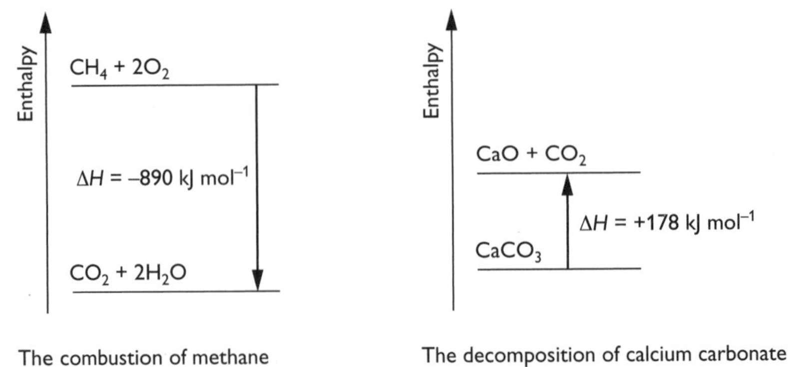

The combustion of methane — **exothermic**

The decomposition of calcium carbonate — **endothermic**

The *change* in enthalpy can be measured (directly for some reactions, or using *cycles* — see below). It is given the symbol **Δ*H***.

- When enthalpy is *lost*, Δ*H* is *negative* and heat is *given out* — an **exothermic reaction**.
- When enthalpy is *gained*, Δ*H* is *positive* and heat is *taken in* — an **endothermic reaction**.

Measuring enthalpy changes directly

Combustion of liquid fuels

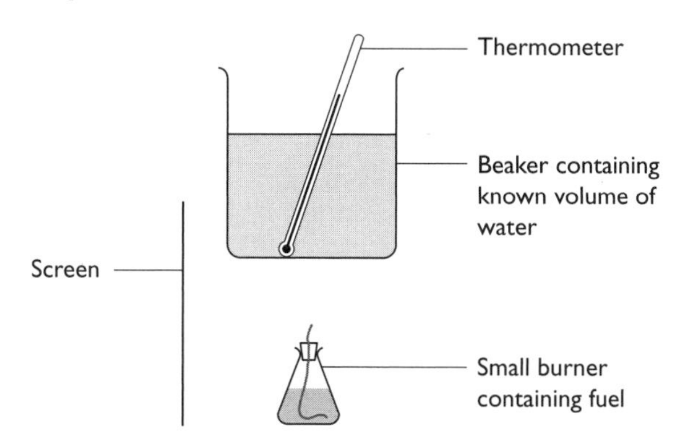

The temperature rise of a known mass of water is measured for a known mass of fuel burnt. The number of kilojoules given to the water can be worked out, using the equation:

heat transferred = **mass (of water)** × **specific heat capacity** × **temperature change (of water)**

Tip You should learn this equation.

You will be told that the specific heat capacity of water is 4.18 (or 4.2) $J\,g^{-1}\,K^{-1}$. This is the amount of energy that is needed to raise the temperature of 1 g of water by 1 K.

Note that:

- The mass of water must be measured in grams (and this is the same as the volume of water in cm^3).
- The temperature change must be measured in K (the same number as when measured in °C, as it is a change).
- The answer will be in joules (divide by 1000 to get kJ).
- You need to work out the number of moles of liquid burnt, using:

$$\textbf{moles} = \frac{\textbf{mass (g)}}{M_r}$$

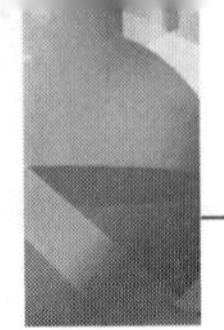

- You should then divide the heat transferred by the number of moles, in order to determine the heat transferred per mole.
- You must not forget the sign (negative here) when you quote this as the enthalpy change of combustion (in $kJ\,mol^{-1}$).

Such calculations are inaccurate because of a number of factors. Some can be minimised, but not completely removed.

Cause of the inaccuracy	Way of minimising the inaccuracy
Heat losses from the apparatus (heat transferred to the surroundings)	• Cover the top of the beaker • Use more screens • Insulate the sides of the beaker
Heat losses to the beaker	• Use a metal calorimeter (metal has a lower specific heat capacity than glass)
Evaporation of the fuel, especially when weighing the burner	• Place a cap on the burner

There may also be uncertainties in the measurements of apparatus.

Quantity	Possible reading	Apparatus	Percentage uncertainty
Temperature change	20°C	Thermometer — two readings to ±0.5°C	1/20 = 5%
Volume of water	200 cm^3	Measuring cylinder ±2 cm^3	2/200 = 1%
Mass of liquid	1 g	Balance — two readings ±0.01 g	0.02/1 = 2%

Thus, it would be worth using a 0.1°C thermometer to reduce the uncertainty.

However, if the result is more than 10% away from the data book value, it is likely to be the case that procedural inaccuracies dominate (as above).

Reactions in solution

Volumes of solutions (at the same starting temperature), or a solution and a solid, can be mixed in a plastic cup and the temperature rise measured.

Again, apply the equation:

heat transferred = mass × specific heat capacity × temperature change

It is usual to make the following simplifying assumptions:

- The mass of water (g) is taken as the volume of the solutions (cm^3).
- The specific heat capacity is taken as that of water (and will be given to you).

(The inaccuracies in these two values tend to cancel each other out when they are multiplied.)

- The energy transferred to any solids that are formed, or to the plastic cup, can be ignored, as these solids have a much smaller mass and specific heat capacity than water.

The calculation is continued by:

- working out the kilojoules transferred
- working out the moles of the reagent that is limiting (that is, not in excess — this will normally be obvious from the question)
- dividing the kilojoules by the moles and inserting the correct sign to get the enthalpy change of the reaction in $kJ\,mol^{-1}$ (*minus* if exothermic, and *plus* if endothermic; note that the '+' must not be omitted)

Heat losses can be minimised by plotting a cooling curve and then extrapolating back to get the highest temperature reached. (This is done in activity DF 2.2.) The plastic cup is quite a good insulator, but it could also be lagged.

Typical uncertainties in the measurements are shown in the table below:

Quantity	Possible reading	Apparatus	Percentage uncertainty
Temperature change	10°C	Thermometer — two readings to ±0.5°C (if a data-logging probe is used, look at the accuracy with which the temperature can be read off the graph)	1/10 = 10% (or less)
Volumes of solution	25 cm^3 each	Pipette ±0.06 cm^3	0.06/25 ≈ 0.2% for each solution

Standard enthalpy changes

Enthalpy changes vary with the conditions. **Standard conditions** are defined as:

- a *specified* temperature, usually 298 K
- one atmosphere pressure
- a concentration of 1 $mol\,dm^{-3}$

A *standard* enthalpy change (with the specified temperature shown) is written $\Delta H^{\ominus}_{298}$.

Standard enthalpy change of combustion ($\Delta H^{\ominus}_c$) is the enthalpy change that occurs when *one mole* of fuel is completely burnt in oxygen under *standard conditions*.

Standard enthalpy change of formation ($\Delta H^{\ominus}_f$) is the enthalpy change that occurs when *one mole* of a compound (in its *standard state*) is formed from its elements (in their *standard states*).

Hess cycles

Hess's law states that the *enthalpy change in going from reactants to products is independent of the route taken*. This can be used to measure enthalpy changes of reaction *indirectly*. Two examples are shown below.

Example 1

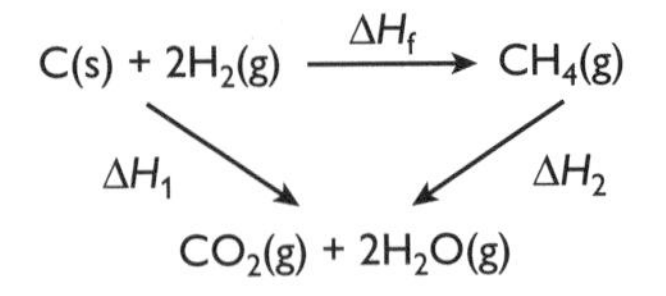

Here ΔH_f is the enthalpy change of formation of CH_4, which cannot be measured directly. ΔH_1 is the sum of the enthalpy changes of combustion of one mole of carbon and two moles of hydrogen. ΔH_2 is the enthalpy change of combustion of one mole of methane.

By Hess's law:

$$\Delta H_f = \Delta H_1 - \Delta H_2$$

(Note: it is *minus* ΔH_2, since we are going *against* the arrow.)

Example 2

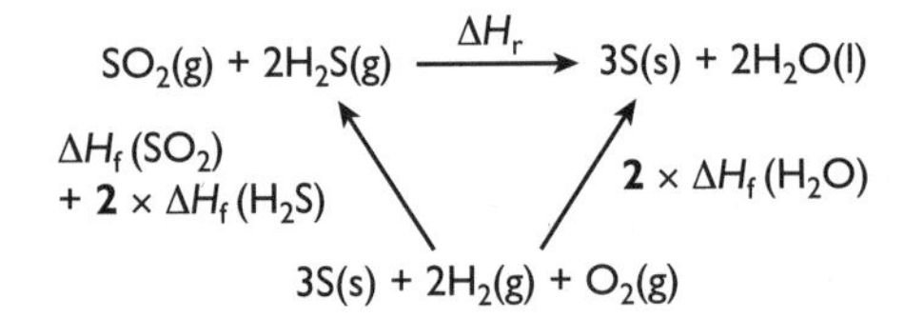

This is how the enthalpy change of a reaction can be measured using enthalpy changes of formation (which are available in tables).

$$\Delta H_r = -(\Delta H_f(SO_2(g)) + 2\Delta H_f(H_2S(g)) + 2\Delta H_f(H_2O(l))$$

Note:

- ΔH_f of an *element* is zero.
- It is vital to *multiply* the ΔH_f values of compounds by the number of moles in the equation.

Often a cycle is not asked for. In this case, the problem in Example 2 can be dealt with by a quicker method, using the equation:

$$\mathbf{\Delta H_r = \text{sum of } \Delta H_f \text{ (products)} - \text{sum of } \Delta H_f \text{ (reactants)}}$$

In Example 2, the ΔH_f values are as follows:

$SO_2(g)$	+ $2H_2S(g)$	→ $3S(s)$	+ $2H_2O(l)$	
−297	2 × −21	0	2 × −286	kJ mol^{-1}

So:

$$\mathbf{\Delta H_r = 2 \times -286 - (-297 + (2 \times -21)) = -233\ kJ\ mol^{-1}}$$

Tip There have been many exam questions requiring use of this equation. Practise these calculations.

Bond enthalpies

The **bond enthalpy** measures the energy required to break a bond into single atoms. It goes against intuition, but when bonds are formed, energy is *given out* so:

- bond-breaking is *endothermic*
- bond-making is *exothermic*

Thus *bond enthalpies* have *positive* signs.

The larger a bond enthalpy, the stronger the bond. Between the same atoms, double bonds are stronger and shorter than single bonds.

A Hess cycle can be drawn that shows how the enthalpy change of a reaction can be determined from bond enthalpies. For example, for the combustion of ethanol:

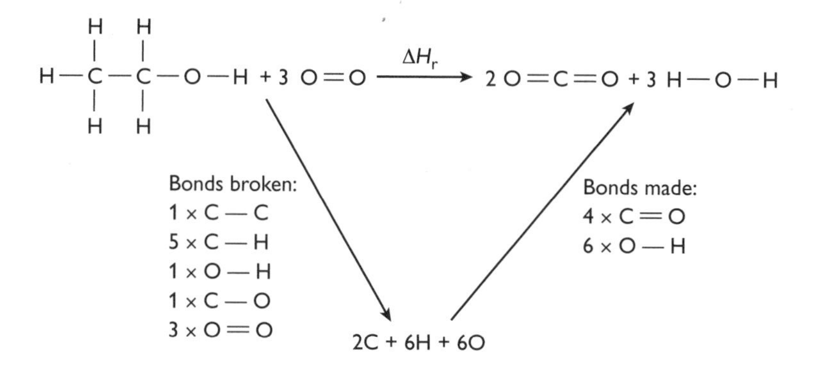

By Hess's law:

$$\Delta H_r = \text{bonds broken} - \text{bonds made}$$

Tip There have been many exam questions requiring use of this equation. Practise these calculations.

Note that:

- The number of C–C bonds broken is *one less* than the number of carbon atoms in the chain.
- Don't forget the oxygen double bond that is broken. For example, if the equation has 2.5 oxygen molecules, multiply 2.5 by the bond enthalpy of O=O.
- It is easy to forget to double the numbers of moles when working out how many C=O and O–H bonds are made, for example wrongly using 2 and 3 (not 4 and 6) here.

Bond enthalpy calculations are useful, but they only give approximate answers because:

- they use *average* values of bond enthalpy for bonds in different compounds
- they rely on all reactants and products being *gaseous* (which is not always the case when they are in their standard states)

Questions & Answers

In this section of the guide there are four questions that test every lettered statement in the specification. They represent the kinds of question which you will get in the unit test, in that they start with a context and contain a wide range of subject matter from the whole specification. Unlike the real thing, there are no lines or spaces left for the answers. Instead, the number of lines is indicated (or a space), e.g. '*(2 lines)*'. The number of marks is, of course, also shown. However, taken together, these questions are longer than a single paper, so do not try to do them all in 75 minutes.

After each question, you will find the answers of two candidates — Candidate A and Candidate B (using different candidates for each question). In each case, Candidate A is performing at the C/D level, while Candidate B is an A-grade candidate.

Examiner comments

All candidate responses are followed by examiner comments. These are preceded by the icon *e* and indicate where credit is due. In the weaker answers, they also point out areas for improvement, specific problems and common errors.

How to use this section

- Attempt to answer the question, giving yourself a time limit of about 1.25 minutes a mark. Do not look at the candidates' answers or examiner comments before you have tried the question yourself.
- Compare your answers with the candidates' answers and decide what the correct answer is; still do not look at the examiner comments while doing this.
- Finally, look at the examiner comments.

Completing this section will teach you a lot of chemistry and improve your exam technique.

Cosmic rays and carbon dating

'Cosmic rays' from space consist of fast-moving electrons. These hit atoms in the upper atmosphere and cause them to emit high-energy neutrons. These neutrons react with nitrogen atoms to form protons and radioactive carbon-14 (which is used for 'radiocarbon dating').

(a) From the particles proton, neutron, electron, choose *two* in each case, which:
- **(i) have the same mass *(1 line)*** (1 mark)
- **(ii) have a charge *(1 line)*** (1 mark)

(b) Use nuclear symbols to complete the equation for the reaction of a neutron with a nitrogen atom to form carbon-14 and a proton:

$^{14}_{7}N + ^{1}_{0}n \rightarrow$ ____________________ (2 marks)

(c) The common isotope of carbon is carbon-12. Compare the nuclei $^{14}_{6}C$ and $^{12}_{6}C$ in terms of their nuclear particles.
- **(i) State (in terms of the numbers of particles they contain) what they both have in common. *(1 line)*** (1 mark)
- **(ii) State (in terms of the numbers of particles they contain) how they differ. *(1 line)*** (1 mark)

(d) (i) What name is given to the number 14 in $^{14}_{6}C$? *(1 line)* (1 mark)
- **(ii) What name is given to unstable isotopes such as $^{14}_{6}C$? *(1 line)*** (1 mark)

(e) The isotope $^{14}_{6}C$ decays with the loss of a β-particle. Write a nuclear equation for this process. *(space)* (3 marks)

(f) Some other isotopes decay giving off an α-particle.
- **(i) Give the nuclear symbol for an α-particle. *(1 line)*** (1 mark)
- **(ii) Describe the difference in penetrating power between α-particles and β-particles. *(3 lines)*** (2 marks)

(g) Geiger and Marsden used α-particles to bombard gold foil. Most of the α-particles passed through undetected, but some bounced back. Before this experiment, the atom had been thought of as a sphere of positive charge dotted with electrons. Explain how Geiger and Marsden's experiment led to a new model of the atom. *(3 lines)* (2 marks)

(h) While living creatures are alive, they exchange carbon with the carbon dioxide in the air. When a creature dies, this exchange stops and the carbon-14 decays. The half-life of the decay process (5370 years) enables the age of a once-living sample to be measured. The ratio of carbon-12 to carbon-14 is $10^{12}:1$ in living things.
- **(i) Suggest one way in which we exchange carbon with our surroundings. *(1 line)*** (1 mark)
- **(ii) Suggest a reason why the presence of carbon-14 is not used as a tracer to follow the path of carbon round the body. *(2 lines)*** (1 mark)
- **(iii) Say what you understand by the term half-life. *(2 lines)*** (1 mark)
- **(iv) In a specimen of wood, three-quarters of the carbon-14 had decayed. How old would the specimen be? *(space)*** (1 mark)

(i) Carbon also forms a stable isotope carbon-13. The presence of this isotope can be detected by a mass spectrometer.

(i) Label the diagram of the mass spectrometer below, writing your labels in the three boxes given. The labels should each describe what is happening to the particles at that point. (3 marks)

(ii) Explain why a carbon-12 ion will pass through the machine more quickly than a carbon-13 ion. ***(2 lines)*** (2 marks)

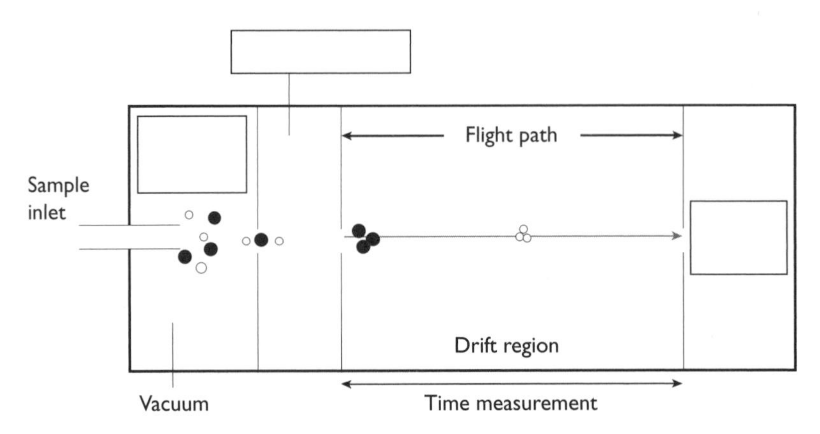

(j) Naturally occurring carbon dioxide contains 1% carbon-13, the rest being carbon-12.

(i) Calculate the A_r of carbon in the sample to two decimal places. ***(space)*** (2 marks)

(ii) Calculate the relative molecular mass of naturally occurring carbon dioxide to two decimal places. (A_r: O, 16.00) ***(space)*** (1 mark)

(iii) The mass spectrum of carbon dioxide shows a peak at m/z = 44. Which ion corresponds to this peak? ***(1 line)*** (1 mark)

(iv) There are other peaks of lower m/z in the mass spectrum of carbon dioxide. How are these peaks formed? ***(1 line)*** (1 mark)

(k) A long-term store of carbon on this planet is calcium carbonate ($CaCO_3$) in rocks. When this is heated it gives off carbon dioxide and leaves calcium oxide.

(i) Write a balanced chemical equation with ***state symbols*** **for the thermal decomposition of calcium carbonate.** ***(space)*** (2 marks)

(ii) Calculate the percentage by mass of carbon in calcium carbonate. ***(space)*** (2 marks)

Total: 34 marks

Candidates' answers to Question 1

Candidate A

(a) (i) Proton and neutron

(ii) Proton only

Candidate B

(a) (i) Proton and neutron

(ii) Proton and electron

Candidate B has got it right and scores two marks. Candidate A scores 1 mark for part (i), but has become muddled in part (ii) and does not score any more. The question would never have been asked in this way if there were only one answer, so it is better to give two particles, even if you are not sure.

Candidate A

(b) $^{14}_{7}N + ^{1}_{0}n \rightarrow ^{14}_{6}C + ^{1}_{1}H$

Candidate B

(b) $^{14}_{7}N + ^{1}_{0}n \rightarrow ^{14}_{6}C + ^{1}_{1}p$

Both candidates have given correct answers. Either 'H' or 'p' is acceptable for a proton in these equations.

Candidate A

(c) (i) Same number of protons (6)

(ii) Different number of neutrons

Candidate B

(c) (i) Six protons

(ii) It has two more neutrons

Candidate A gets part (i) right but in part (ii) is being vague. *State in terms of numbers of particles* says the question, so he should have given the numbers of protons. He only scores 1 mark for part (i). Candidate B gives the correct detail in (i) and scores 1 mark. It is not clear which isotope she means by 'it' in (ii), so she does not score the mark.

Candidate A

(d) (i) Relative atomic mass

Candidate B

(d) (i) Mass number or nucleon number

Candidate A is wrong. Relative atomic masses are not necessarily whole numbers. The top number in the nuclear symbol is the mass number, as stated by Candidate B. She does not lose marks for giving two alternative answers as both are correct, but it is best to give just one answer.

Candidate A

(d) (ii) Unstable

Candidate B

(d) (ii) Radioactive

Candidate A's answer is not wrong but it gains no credit as the word 'unstable' is used in the question. Candidate B is correct.

Candidate A

(e) $^{14}_{6}C \rightarrow {}^{14}_{5}B + {}^{0}_{-1}e$

Candidate B

(e) $^{14}_{6}C \rightarrow {}^{14}_{7}N + e^-$

e Candidate A has made a fairly common error. Five minus one does not equal six on the bottom row. However, the symbol for the electron is correct and the atomic number 5 does correspond to boron, so he will score 2 out of 3 marks.

Two marks also for Candidate B, who, for some strange reason, has not given the full nuclear symbol for the electron (Candidate A's is correct).

Candidate A

(f) (i) He

(ii) α-particles do not penetrate far and β-particles penetrate further.

Candidate B

(f) (i) $^{4}_{2}He$

(ii) α-particles will hardly penetrate through paper but β-particles penetrate thin sheets of metal.

e Candidate A has not written a *nuclear symbol* for the α-particle, so will not score for part (i). In part (ii), he has not given nearly enough detail. He should have seen there were 2 marks, also the word 'describe' rather than 'state' should have given him a clue. He would just score 1 of the 2 marks here for realising that α-particles do not penetrate as far as β-particles.

Candidate B scores 1 mark in part (i). She also scores 2 marks in part (ii) for giving details for both particles. 'Thin sheets of metal' is not an ideal answer and it would have been safer to add '...such as aluminium'.

Candidate A

(g) It showed that most of the atom was empty space.

Candidate B

(g) The dense, small nucleus deflects a very few α-particles.

e Both candidates score 1 out of 2 marks. Candidate A is not precise but just scores the mark. He should have added that the 'empty space' contained electrons which did not deflect the α-particles. Candidate B has written a good answer for part of the question, but has not explained why most of the α-particles are undeflected.

Candidate A

(h) (i) Breathing out

(ii) It is everywhere.

Candidate B

(h) (i) Eating food

(ii) Its half-life is so long that it would not give enough radiation.

e This is an 'application of knowledge' question, as the word 'suggest' shows. Candidate A has given a weak answer to part (i). A much better answer would be 'respiration', but Candidate A would get the mark, as this is not in the specification. (If in doubt, write something sensible.) Candidate B's answer is also correct. There are many possible answers to part (ii), all of which require understanding of what a *tracer* means. Candidate A's answer is just acceptable as it implies that, since carbon-14 is in all life, it cannot be used to follow the passage of carbon. Candidate B's answer is better. Another possible answer is that the amounts of carbon-14 are too small. Both candidates score full marks in part (h). *Note that the 'stem' to part (h) gives many clues.*

Candidate A

(h) (iii) This is the time when the radioactivity has fallen to half its initial value.

Candidate B

(h) (iii) Half-life is the time taken for half the radioactive atoms to decay.

e Each candidate scores the mark. Candidate A is right because the amount of radioactivity is proportional to the number of radioactive atoms remaining.

Candidate A

(h) (iv) 2685 years

Candidate B

(h) (iv) Half decays in the first half-life. Then a further quarter decays in the second half-life. So the sample is two half-lives old: 10 740 years

e Candidate A has presumably misread the questions as 'three quarters of the C-14 remain'. Thus the answer is wrong and does not score the mark. (Even if the question had said this, the answer would not be right, as the decay process is not linear.) Candidate B has argued the answer well and deserves the mark.

Candidate A

(i) (i)

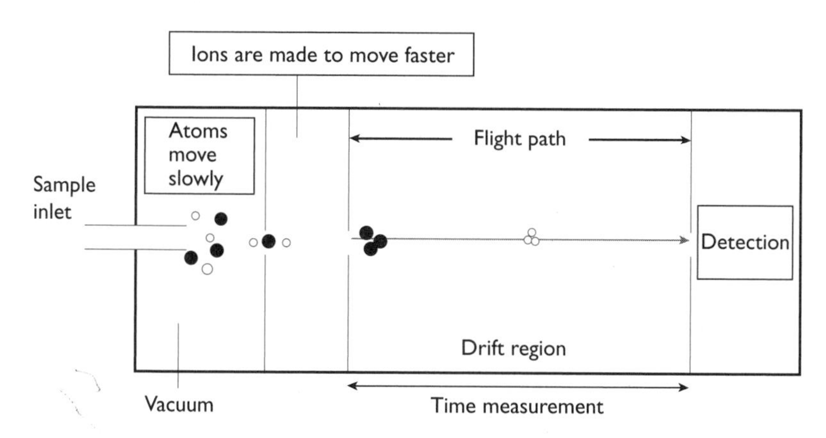

Candidate B

(i) (i)

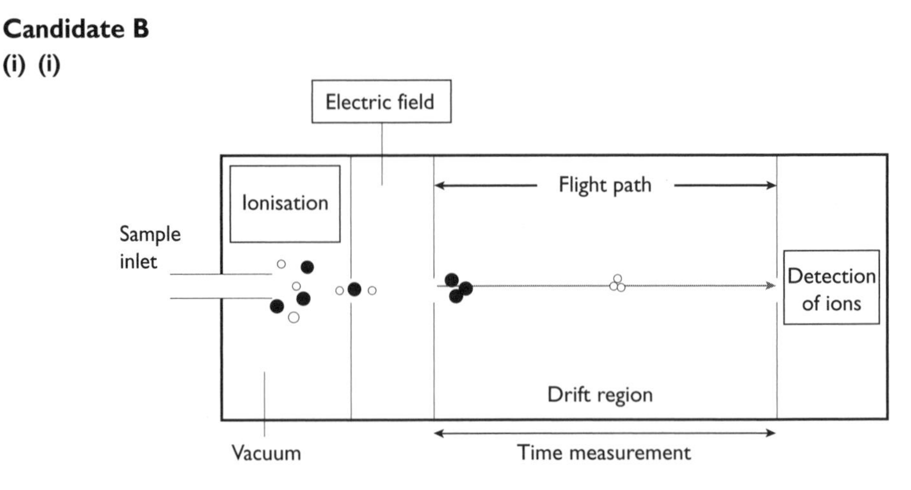

The labels should read 'ionisation', 'acceleration (of ions)' and 'detection (of ions)'. Candidate A has the first wrong, though the second and third score. So he gets 2 out of 3 marks.

Candidate B also gets 2 out of 3 marks. 'Electric field' is what is present, but it does not answer the question, as it does not explain what is happening to the ions.

Candidate A

(i) (ii) It has smaller mass, therefore greater velocity.

Candidate B

(i) (ii) Both particles have the same kinetic energy, so carbon-12, with the smaller mass, will have the greater velocity.

Candidate A starts with 'it', which is unwise. However, he gets away with this word as the question is about carbon-12. He leaves out the point about all the ions starting with the same kinetic energy, and thus only scores 1 out of 2 marks.

Candidate B makes both points and scores 2 marks.

Candidate A

(j) (i) 1% of 12 is 0.12. So the RAM is 12.12

Candidate B

(j) (i)

$$A_r\,C = \frac{(99 \times 12) + (1 \times 13)}{100} = 12.01\text{ g}$$

Candidate A has gone off on the wrong track. He has tried a short-cut and failed. He scores nothing.

Candidate B has used the expression that she has learnt, and has obtained the correct answer of 12.01. However, she has lost the second mark by putting the unit 'g'. Relative atomic masses have no units. She scores 1 mark.

Candidate A

(j) (ii) 12.12 + (2 × 16.00) = 44.12

Candidate B

(j) (ii) 12.01 + 2 × 16 = 44.01

Candidate A would score 1 mark by 'error carried forward' from part (i), using the wrong value of A_r for carbon. Candidate B has done it right (with the correct A_r) and scores 1 mark.

Candidate A

(j) (iii) The M^+ ion.

Candidate B

(j) (iii) The ion which is a CO_2 molecule with one electron knocked off.

Both answers are different ways of saying the same thing, so both score.

Candidate A

(j) (iv) Bits are knocked off the molecule.

Candidate B

(j) (iv) These are fragments of the molecule.

Candidate B is correct and scores the mark. Candidate A has just about implied the same thing and scores also.

Candidate A

(k) (i) $CaCO_3(s)$ + heat → $CaO(s)$ + $CO_2(g)$

Candidate B

(k) (i) $CaCO_3(s)$ → $CaO(s)$ + $CO_2(g)$

Candidate A should realise that 'heat' is not a chemical reagent and should not be written into equations. He would not score the first mark because of this, but he scores the second one for the state symbols.

Candidate B has produced the correct answer and scores 2 marks.

Candidate A

(k) (ii) $CaCO_3$ = 40 + (3 × 28) = 112 % = 12/112 = 10.7%

Candidate B

(k) (ii) M_r $CaCO_3$ = 100 12/100 = 0.12 = 12%

Candidate A has made a careless error again and assumed the formula is $Ca(CO)_3$. The rest of the working is right, though rather abbreviated, so he would score 1 mark.

Candidate B has got it right (though she would have been well advised to use 100.1 as the M_r of $CaCO_3$, as the A_r of calcium is shown on the periodic table as 40.1). She scores 2 marks.

Candidate A has scored 19 marks and Candidate B 28 marks.

Hydrogen in the universe

The commonest element in the universe is hydrogen. The absorption spectrum of the gases surrounding the Sun shows that these gases contain hydrogen.

(a) An absorption spectrum consists of dark lines on a coloured background.

(i) Explain how hydrogen atoms can absorb light. ***(2 lines)*** (2 marks)

(ii) Give one way in which the ***emission*** **spectrum for hydrogen would be the same as its absorption spectrum.** ***(2 lines)*** (1 mark)

(iii) Give one way in which the emission spectrum would differ from the absorption spectrum. ***(2 lines)*** (1 mark)

(iv) Suggest how the discovery and explanation of atomic spectra contributed to the development of the model of the atom. ***(2 lines)*** (1 mark)

(b) All the other elements in the universe are made from hydrogen by nuclear reactions. What name is given to the type of nuclear reaction involved? ***(1 line)*** (1 mark)

(c) Hydrogen forms two ions, one of which is H^-.
How many electrons are there in this ion? ***(1 line)*** (1 mark)

(d) In modern periodic tables, hydrogen is sometimes placed above group 1. The main reason for classifying hydrogen with group 1 is that it forms H^+ ions.

(i) Why is hydrogen placed first in a modern periodic table? ***(2 lines)*** (1 mark)

(ii) Unlike hydrogen, lithium and all the other group 1 metals show metallic bonding. Draw a labelled diagram which illustrates metallic bonding. ***(space)*** (2 marks)

(e) Sometimes, hydrogen is placed above group 7 in the periodic table.

(i) Suggest ***two*** **similar ways in which hydrogen and fluorine form chemical bonds with other elements.** ***(3 lines)*** (2 marks)

(ii) Mendeleev reversed the positions of Te and I in his periodic table based on relative atomic masses, so that iodine was in group 7.

- **Why did he do this?**
- **Which later discovery validated this reversal of the elements?** ***(3 lines)*** (2 marks)

(f) (i) Hydrogen forms a hydride with magnesium in which 4.86 g of magnesium combine with hydrogen to form 5.26 g of the hydride. Calculate the formula of magnesium hydride. ***(space)*** (2 marks)

(ii) Write the electron structure of magnesium (for example, that for nitrogen is 2.5).
Explain how you deduced this from the periodic table. ***(3 lines)*** (2 marks)

(iii) Predict the charge on a magnesium ion, giving your reasoning. ***(2 lines)*** (2 marks)

(g) Magnesium reacts with oxygen to form a white ionic solid, MgO, that is only slightly soluble in water.

(i) Predict two other physical properties of magnesium oxide that arise from its ionic structure. ***(2 lines)*** (2 marks)

(ii) Draw a '*dot-and-cross*' diagram (showing outer electron shells only) for the oxide ion O^{2-}. ***(space)*** (1 mark)

(iii) Magnesium oxide reacts slightly with water to form magnesium hydroxide. Suggest the pH of magnesium hydroxide solution, giving a reason for this. ***(3 lines)*** (2 marks)

Total: 25 marks

■ ■ ■

Candidates' answers to Question 2

Candidate A

(a) (i) Hydrogen atoms absorb light when they are excited to higher energy levels.

Candidate B

(a) (i) When electrons in the hydrogen atom absorb light energy, they rise to higher energy levels.

e Candidate A has left out a vital word — '*electrons*'. However, she has got the idea of moving up energy levels, so she would score 1 out of 2 marks. Candidate B scores both marks.

Candidate A

(a) (ii) The lines would be the same.

Candidate B

(a) (ii) The frequencies are the same.

e Both candidates are correct and score 1 mark.

Candidate A

(a) (iii) The lines would be black.

Candidate B

(a) (iii) The lines would be coloured on a black background.

e Candidate A has described an absorption spectrum, which is mentioned in the stem of the question. She does not score. Candidate B has described an emission spectrum, which just answers the question and scores the mark, though in such comparison questions he would be well advised to say what it is. He scores the mark.

Candidate A

(a) (iv) It showed that the electrons were not just anywhere around the atom, but in energy levels.

Candidate B

(a) (iv) It is evidence for energy levels in the atom.

question

Candidate A is correct here and scores the mark, while Candidate B has left out any reference to electrons and thus does not score.

Candidate A

(b) Addition

Candidate B

(b) Fusion

Candidate A is thinking along the right lines but 'addition' is not used by chemists here and so does not score. '*Fusion*' is the correct term, so Candidate B scores the mark.

Candidate A

(c) 2

Candidate B

(c) None

Candidate A is right. There is one electron in a hydrogen atom, and another is added to make the 1^- ion. Candidate B is presumably confusing the H^- ion with H^+, which, indeed, has no electrons. He does not score.

Candidate A

(d) (i) Because it is the smallest atom.

Candidate B

(d) (i) Because the periodic table is arranged in order of atomic number.

Neither candidate scores this easy mark, because they have not been detailed enough. Candidate B is closer, but he needed to say that hydrogen has an atomic number of one to get the mark.

(d) (ii)

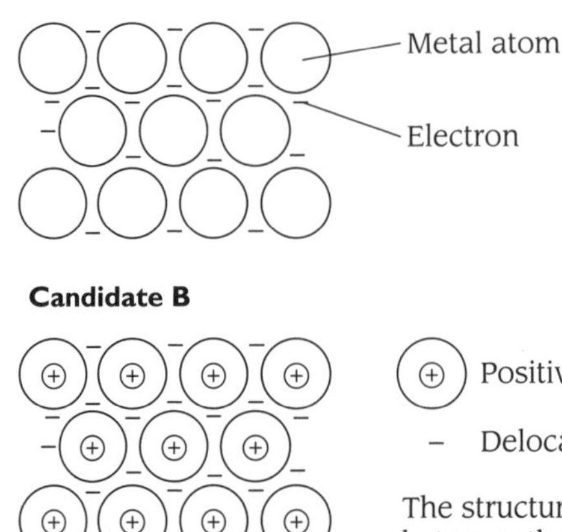

Candidate A has shown a regular lattice of metal particles and the electrons, so she scores one mark. She has not, however, indicated (or labelled) that the metal particles are ions.

Candidate B has given a good answer and gets 2 marks. He has added the way in which the structure is held together, which is correct but not essential here.

Candidate A

(e) (i) Hydrogen and fluorine form a single covalent bond together by sharing one pair of electrons.

Candidate B

(e) (i) Either single covalent bonds or 1– ions.

Candidate A has not read the question to its conclusion. It does not ask how hydrogen and fluorine combine with each other but how they combine with *other elements*. However, she would score 1 mark for talking about single covalent bonds. Candidate B has answered the question fully, and gains 2 marks.

Candidate A

(e) (ii) Their properties fit better this way round. Atomic number was discovered.

Candidate B

(e) (ii) Iodine fits better in group 7 and tellurium in group 6.
The discovery of atomic number, as Te and I are the 'right' way round when arranged by atomic number.

Both score two marks. Candidate A's answers might have caused the examiner to check the match with the mark-scheme, as they are brief. However, they just fit. Beware of such short answers, though, as they might not always be considered to have enough detail to score. Candidate B has not mentioned the word 'properties', and thus only just scores the mark. However, he has given a good answer to the second bullet-point, which is the type of answer you should always aim to give.

Candidate A

(f) (i) Moles magnesium = 4.86/24 = 0.2

Moles hydrogen = 5.26/1 = 5.26.

Formula = MgH_{26}

Candidate B

(f) (i) Moles Mg = 4.86/24.3 = 0.2

Mass H = moles H = 5.26 – 4.86 = 0.4

Formula is MgH_2

Candidate A has failed to spot that the second mass is that of the *hydride*, not hydrogen. Still, she has worked out correctly the ratio of the moles that she has calculated, and she scores 1 mark by 'error carried forward', though she ought to realise that this is an unlikely formula.

question

Candidate B has done the calculation correctly and scores 2 marks. He has used 24.3 as the A_r of magnesium, which makes the working easier. Remember to look at the values of A_r in the periodic table on the *Data Sheet* and use the values to one decimal place.

Candidate A

(f) (ii) 2.8.2. Magnesium is in group 2 and thus has two electrons in its outer shell.

Candidate B

(f) (ii) $1s^2 2p^6 3s^2$. Magnesium is in period 3 and group 2.

Candidate A scores for the electron structure. However, her explanation does not distinguish magnesium from all the other group 2 elements, and so would not score the second mark.

Candidate B does not answer the question as set. He attempts the electron configuration in terms of s- and *p*-electrons and gets it wrong. (If he had got it right [$1s^2 2s^2 2p^6 3s^2$], under the mark-scheme used he would have scored, but this is risky.) His explanation is adequate and would score. Mention of the fact that magnesium has an atomic number of ten would be helpful, too. So he scores 1 mark.

Candidate A

(f) (iii) 2+. Magnesium is in group 2.

Candidate B

(f) (iii) 2+. Magnesium has two electrons in its outer shell that are easily lost.

Candidate A scores 1 mark for the charge. The reasoning is on the right lines, but not good enough to score. The mark scheme here required reference to electrons (partly because the group number is an answer to an earlier question). Candidate B mentions electrons and scores both marks.

Candidate A

(g) (i) Soluble in water and conducts electricity

Candidate B

(g) (i) High melting point and conducts when molten

Candidate A has failed to read the question when writing her first response. The question asks for *other* physical properties, and solubility in water is mentioned in the stem. On the second point, she is too vague. Magnesium oxide only conducts electricity when it is molten (when the ions can move). She scores no marks here.

Candidate B scores both marks. He has omitted to mention conducting *electricity* in the second part, but he gets away with this under the mark-scheme.

(g) (ii)

Candidate A	**Candidate B**
:Ö: (with two dots below)	[×× Ö :]$^{2-}$ (with two dots below)

Candidate A has the idea that there are eight electrons round the oxygen, but this is not enough to score the mark. The charge on the ion is essential. Candidate B's answer is correct and scores the mark. The inclusion of a pair of crosses, replacing one pair of dots, is not essential, but it indicates that two of the electrons were added to the oxygen atom to make the oxide ion.

Candidate A

(g) (iii) Group 2 hydroxides are alkaline in solution.

Candidate B

(g) (iii) pH 14; magnesium hydroxide is basic.

Candidate A just scores both marks, as 'alkaline' is just acceptable for an answer that should be a pH number, and the reasoning is good. 'Alkaline' is acceptable, as the pH number allowed would be anything from eight to 14, as it depends on the solubility of the magnesium hydroxide and its concentration.

Candidate B scores the first mark, but not the second. Basic hydroxides are not always soluble in water, though this is a small quibble. He fails to score because he doesn't add anything to the first part. An answer like Candidate A's is required to give the periodic-table context for the prediction.

Candidate A has scored 13 marks, and Candidate B 20 marks.

Isomers in petrol

Three hydrocarbons which are found in car petrol are pentane and its two structural isomers. These contribute different volatilities and octane numbers to the fuel.

(a) (i) Name a naturally occurring source of hydrocarbons such as pentane. ***(1 line)*** (1 mark)

(ii) To which homologous series do pentane and its isomers belong? ***(1 line)*** (1 mark)

(iii) Draw the full ***structural formula*** **of pentane (C_5H_{12}).** ***(space)*** (1 mark)

(iv) Draw ***skeletal*** **formulae for the two** ***isomers*** **of pentane, and name them.** ***(space)*** (4 marks)

(b) (i) State how a fuel's ***octane number*** **affects how it performs in an engine.** ***(2 lines)*** (1 mark)

(ii) Which one out of pentane and its two isomers would have the highest octane number? Give a reason for your answer. ***(2 lines)*** (2 marks)

(iii) Name a reaction of a hydrocarbon (apart from ***isomerisation*****) which will improve the octane number.** ***(1 line)*** (1 mark)

(c) (i) Write a balanced chemical equation for the complete combustion of pentane. ***(space)*** (2 marks)

(ii) Use your equation and the bond enthalpy data below to calculate a value for the enthalpy change of combustion of pentane. ***(space)*** (3 marks)

Bond	Bond enthalpy/kJ mol^{-1}
C–C	+347
C–H	+413
O=O	+498
C=O	+805
O–H	+464

(iii) The enthalpy changes of combustion of the two isomers of pentane are similar to your calculated value. Suggest why. ***(3 lines)*** (2 marks)

(iv) Which of the bonds in the table is the strongest? Give a reason for your answer. ***(2 lines)*** (1 mark)

(d) An engine running on pentane would produce carbon monoxide and nitrogen monoxide. Describe how these pollutants arise in the engine and describe their polluting effects. ***(8 lines)*** (5 marks)

(e) Carbon monoxide and nitrogen monoxide can be removed from the car exhaust by a catalytic converter containing a platinum/rhodium catalyst on a ceramic support.

(i) What do you understand by the term ***catalyst*****?** ***(2 lines)*** (1 mark)

(ii) Explain why this catalyst is described as ***heterogeneous*****.** ***(2 lines)*** (1 mark)

(iii) Write a balanced chemical equation for the reaction of carbon monoxide with nitrogen monoxide to produce two less harmful gases. ***(space)*** (2 marks)

(iv) Explain in outline how carbon monoxide and nitrogen monoxide react on the surface of the heterogeneous catalyst. ***(space and 4 lines)*** (3 marks)

In your answer, you should use appropriate technical terms, spelt correctly.

(v) Lead compounds in the petrol would poison the catalyst. Explain what *catalyst poison* means. (*2 lines*) (2 marks)

Total: 33 marks

■ ■ ■

Candidates' answers to Question 3

Candidate A

(a) (i) Crude oil

Candidate B

(a) (i) Fractional distillation of crude oil

Both candidates are right and score the mark. Candidate B's addition of 'fractional distillation' is correct but not essential.

Candidate A

(a) (ii) Hydrocarbons

Candidate B

(a) (ii) Alkanes

The isomers are *hydrocarbons*, but so is any substance containing just carbon and hydrogen. This is too wide a statement, so Candidate A does not score. The correct answer is *alkanes*, so Candidate B scores the mark. Learn to recognise the homologous series you need for this unit: alkanes, cycloalkanes, alkenes, arenes, alcohols and ethers.

Candidate A

(a) (iii)

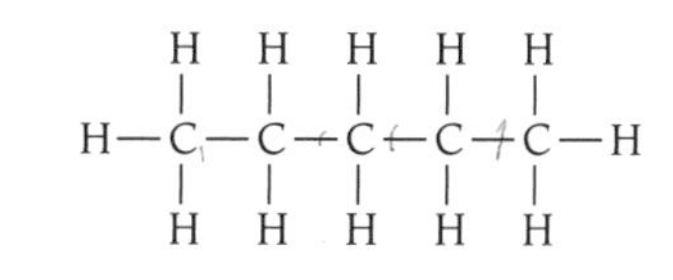

Candidate B

(a) (iii) $H_3C–CH_2–CH_2–CH_2–CH_3$

Candidate A is correct here and scores the mark. He has drawn a full structural formula showing all the bonds and atoms. Candidate B has drawn a shortened structural formula which would not gain any credit. It is a correct representation of pentane but it *doesn't answer the question*.

Candidate A

(a) (iv)

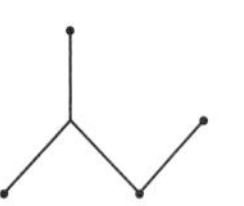

2-methyl butane

3-methylbutane

Candidate B

(a) (iv)

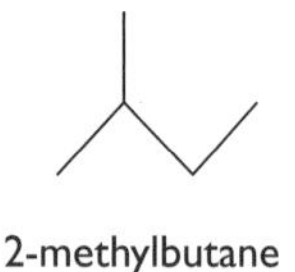

2-methylbutane

2,2-methylpropane

Candidate A has got the right idea, but he has made many mistakes. First, do not put dots on skeletal structures. Second, these two are the same structure. Third, there should not be a gap between methyl and butane in the name. He would lose one of the skeletal formula marks for putting on dots (and he won't score the second one anyway). He would score just 1 mark for the left-hand name, as the examiner would not penalise the gap.

What about Candidate B? She has correctly identified the two structures and drawn correct skeletal formulae for them. However, she has made a mistake in the right-hand name. Although the '2,2' shows there are two groups on the second carbon, it is also necessary to give 'di'. Thus the correct name is 2,2-*di*methylpropane. Thus, she would score 3 out of 4 marks.

Candidate A

(b) (i) The octane number measures the engine's tendency to knock.

Candidate B

(b) (i) The octane number measures how likely the fuel is to cause auto-ignition in the engine. The higher the number, the less likely it is to do so.

Candidate A mentions knocking, which is permissible as an alternative to 'auto-ignition'. However, he fails to say which way round it works (the higher the number, the *less* likely it is to knock) so he would not score. Candidate B describes knocking in terms of auto-ignition, and she also states the effect of a higher octane number, so she would score the mark.

Candidate A

(b) (ii) 2,2-dimethylbutane, because it is the most volatile one.

Candidate B

(b) (ii) 2,2-dimethylbutane, because it is the most branched.

Candidate A has named the correct isomer, but the reason is not valid. Do not attempt to explain octane numbers by volatility (the ease with which a fuel evaporates). They are often related, with the highest octane numbers going with the most volatile fuels, but a fuel does not have a high octane number *because* it is volatile. So Candidate A scores 1 mark.

Candidate B scores two once again. Her answer is exactly right.

Candidate A

(b) (iii) Adding an oxygenate

Candidate B

(b) (iii) Reforming

Candidate A has not read the question, which asks for a *reaction* of a hydrocarbon. Thus, his answer, while correct for a question such as 'how may the octane number of fuels be improved?', scores zero here.

Candidate B was perhaps set on the right track by the word 'isomerisation' in the question. She had learnt this in the context of *reforming* and *cracking*, either of which would have scored the mark.

Candidate A

(c) (i) $C_5H_{12} + 11O_2 \rightarrow 5CO_2 + 6H_2O$

Candidate B

(c) (i) $C_5H_{12} + 8O_2 \rightarrow 5CO_2 + 6H_2O$

Candidate A would score 1 mark for getting the products — carbon dioxide and water — correct. However, he has made a balancing error, just adding the coefficients of CO_2 and H_2O to get the coefficient for oxygen, not realising that there is only one atom of oxygen per molecule of H_2O. Candidate B is right and scores 2 marks.

Candidate A

(c) (ii)

Bonds broken

5 × C–C	1735
12 × C–H	4956
Total	6691

Bonds made

5 × C=O	4025
6 × O–H	2784
Total	6809

$\Delta H = 6691 - 6809 = -118 \text{ kJ mol}^{-1}$

Candidate B

(c) (ii)

Bonds broken

4 × C–C	4 × 347	1388
12 × C–H	12 × 413	4956
8 × O=O	8 × 498	3984
Total		10 328

Bonds made		
10 × C=O	10 × 805	8050
12 × O–H	12 × 464	5568
Total		13 618

$\Delta H = 10\,328 - 13\,618 = -3290\ \text{kJ mol}^{-1}$

Candidate A has made a good attempt here. Many weaker candidates do not give themselves a chance, just jotting down odd figures and arriving at an answer apparently at random. Because Candidate A has laid out the working quite well, it is possible to see the errors he has made and it is possible for the examiner to reward him when he has got things right. He has started by making a common mistake, thinking that there are five C–C bonds in pentane, when, an inspection of his answer to part (b) would show him there are only four. He has also left out the O=O bonds. So, no marks so far. Under bonds made, he has made another common mistake in not noticing that there are *two* C=O bonds in a carbon dioxide molecule and two O–H bonds in a water molecule. Thus he loses the mark for bonds made. Then, however, he writes down the correct working to calculate ΔH (bond broken − bonds made) and calculates it correctly. Even though he should have been surprised at the small size of his answer, he would score the last mark.

Candidate B is on a winning streak. She has given even more detail (useful if you want to go back and double-check, for example) and got the calculation completely right, so she scores 3 marks.

Candidate A

(c) (iii) They are all really the same compound.

Candidate B

(c) (iii) They all have the same number and type of bonds.

This is a difficult part and Candidate A has not grasped the idea. However, he has written something which might be relevant, though it does not score in this case.

Candidate B has scored 1 out of the 2 marks available. She is on track but does not give enough detail (signalled by the 2 marks and the 3 lines). The second mark would be awarded for saying that bond enthalpies do not vary much between compounds.

Candidate A

(c) (iv) C=O, since it has the greatest bond enthalpy.

Candidate B

(c) (iv) C–H, since these bonds are the most unreactive.

Candidate A's answer is right, so he would score the mark. Candidate B's answer is thoughtful but incorrect. The strength of a bond is measured by its bond enthalpy. The reactivity of a bond depends on a number of factors.

Candidate A

(d) When the fuel is not fully combusted, carbon monoxide is produced, which is toxic. Nitrogen monoxide is formed in the engine when nitrogen compounds in petrol are not fully combusted. It leads to the production of acid rain.

Candidate B

(d) Carbon monoxide would be produced from incomplete combustion of the pentane. It is a harmful gas, as is nitrogen monoxide, which is produced when nitrogen from the air reacts with oxygen in the heat of the engine.

There are 5 marks. Candidate A would score 1 mark for the point about the formation of carbon monoxide and another for its polluting effect, though a fuller answer would have said that it stops the blood taking up oxygen. The explanation of the formation of nitrogen monoxide is incorrect, but the polluting effect is correct, so Candidate A scores 3 out of 5 marks.

Candidate B would score 1 mark for the formation of carbon monoxide. She would not score for 'harmful' on its own, 'harmful to life' or 'toxic' is needed. Her answer about the formation of nitrogen monoxide is also good, worth two marks. The question does not disallow writing the same polluting effects for both gases, but 'harmful' would not score here either. So she also gets 3 out of 5 marks.

Candidate A

(e) (i) A catalyst is a substance that speeds up a reaction by lowering its activation energy.

Candidate B

(e) (i) A catalyst speeds up a reaction but does not take part in it.

These are two rather different definitions. Candidate A would score the mark. Candidate B would not score as she has confused 'chemically unchanged at the end' with 'not taking part'. You know that heterogeneous catalysts, for example, take part in the reaction (see below). So the mark scheme requires 'speeds up reactions' for the first point, with one of either 'unchanged at the end' or 'lowers activation enthalpy/energy' for the second point. Suppose someone writes 'Speeds up a reaction by lowering the activation enthalpy, though it does not take part'. This would not score 1 mark, as the incorrect last point cancels out the correct second one. This may seem hard, but examiners do this to stop candidates giving a whole list of answers and encouraging the examiner to pick the right one and ignore the rest. Bear this in mind — don't 'hedge your bets'.

Candidate A

(e) (ii) It is in a different state.

Candidate B

(e) (ii) The reactants and catalyst are in different physical states.

Candidate A is too vague to score, but Candidate B scores with a good answer.

Candidate A

(e) (iii) $CO + NO \rightarrow CO_2 + N$

Candidate B

(e) (iii) $2CO + 2NO \rightarrow 2CO_2 + N_2$

Candidate A would score 1 mark as he has identified the correct products (carbon dioxide and nitrogen) and written a balanced equation for their formation. However, he has made an error by forgetting that nitrogen gas in the air is N_2. Candidate B's equation is completely correct and would score 2 marks.

Candidate A

(e) (iv) The gases are adsorbed on the surface, where they react and the products leave.

Candidate B

(e) (iv)
- The gases are absorbed on to the surface of the catalyst.
- Bonds are weakened and break.
- New bonds are formed.
- The products leave the surface.

The questions says 'outline the stages', so an answer like Candidate B's is more likely to score, as it separates out the processes. Candidate A scores the first mark for 'adsorbed', but then is too vague about bonds to score the second mark. Then he will pick up 1 mark for 'the products leave', scoring 2 marks in all.

Candidate B hasn't quite answered the question, which is about CO and NO, not gases in general. To make it an even better answer, she could have indicated *which* bonds were being made and broken. She has also missed the difference between *absorbed* (a physical process) and *adsorbed* (a chemical binding), so she will have lost the first mark, as she has not 'spelt the technical terms correctly'. She also scores 2 marks.

Candidate A

(e) (v) Lead blocks the active sight of the catalyst and stops it working.

Candidate B

(e) (v) A poison is adsorbed on to the surface of the catalyst, so that the reactant molecules cannot be adsorbed and react themselves.

Candidate A would score 1 mark for 'stops it working'. However, he needs a word more exact than 'blocks' (*adsorbed* or *binds*, for example) and he has confused catalysts with enzymes. A catalyst's surface is active all over, whereas enzymes have *active sites* (note spelling) where reaction occurs.

Candidate B has scored both the mark for 'adsorbed' on to the surface and the other for stopping the reaction, so she scores 2 marks.

Candidate A has scored 15 marks and Candidate B 25 marks.

Uses of ethanol

Ethanol can be used as a *biofuel* that is an alternative to petrol.

(a) (i) The enthalpy change of combustion of liquid ethanol can be measured by burning ethanol to heat water. Draw a labelled diagram of a simple apparatus you could use for this. *(space)* (3 marks)

(ii) In such an experiment, the burning of 2.03 g of ethanol caused the temperature of 500 g of water to rise by 20°C. Calculate the energy transferred to the water, in joules. Specific heat capacity of water = 4.2 $J g^{-1} K^{-1}$ *(space)* (1 mark)

(iii) Calculate the value this gives for the enthalpy change of combustion of ethanol in $kJ\,mol^{-1}$. (M_r for ethanol is 46.) Give your answer to two significant figures. *(space)* (3 marks)

(iv) Give an assumption that you have made in doing this calculation. *(2 lines)* (1 mark)

(v) The uncertainty in the balance used to measure the ethanol is ±0.01 g. The uncertainty in the thermometer is ±0.5°C. Calculate the percentage uncertainties in the mass of ethanol and the temperature rise, given that both involve two measurements. *(2 lines)* (2 marks)

(vi) Suggest one reason (*apart from experimental inaccuracies*) why this method is unlikely to measure the *standard* enthalpy change of combustion of ethanol. *(3 lines)* (1 mark)

(b) The equation for the complete combustion of ethanol is:

$C_2H_5OH(l) + 3O_2(g) \rightarrow 2CO_2(g) + 3H_2O(l)$ Equation 4.1

(i) You are given this equation and the standard enthalpy changes of formation of ethanol, carbon dioxide and water. Construct a Hess cycle which could be used to calculate the value of the standard enthalpy change of combustion of ethanol at 298 K. *(space)* (2 marks)

(ii) Use your cycle and the data given below to calculate the value of the standard enthalpy change of combustion of ethanol. *(space)* (3 marks)

Compound	$\Delta H^{\ominus}_{f,298}$/kJ mol^{-1}
$C_2H_5OH(l)$	−277
$CO_2(g)$	−394
$H_2O(l)$	−286

(c) Other alcohols are also found in petrol. Draw the full structural formula for propan-1-ol. *(space)* (1 mark)

(d) Equation 4.1 also applies when ethanol burns in a car engine.

(i) Calculate the volume of oxygen which would react exactly with 1.0 dm^3 ethanol *vapour* according to Equation 4.1. (Assume that both volumes are measured at the same temperature and pressure.) *(space)* (1 mark)

(ii) Calculate the volume of carbon dioxide (at room temperature and pressure) that would result from burning 2.3 g of ethanol. (M_r for ethanol is 46; 1 mole of a gas occupies 24 dm^3 at room temperature and pressure.) *(space)* (2 marks)

(e) (i) Explain why liquid ethanol has a lower *entropy* than gaseous oxygen. *(2 lines)* (2 marks)

(ii) In the reaction in Equation 4.1, would you expect the entropy to increase, decrease or stay the same? (Assume that all the reactants and products are gases.) Give a reason for your answer. *(2 lines)* (2 marks)

(f) (i) Draw a '*dot-and-cross*' diagram for ethanol. *(space)* (1 mark)

(ii) This '*dot-and-cross*' diagram is a *model* of the ethanol structure. Give one way in which it does not describe the structure of ethanol very well. *(2 lines)* (1 mark)

(iii) State the values of the following bond angles in the ethanol molecule. *(1 line)* (2 marks)

(iv) Explain your answer to f(iii) part (2). *(3 lines)* (3 marks)

(g) State, with a reason, whether you would class ethanol as an *aliphatic* or an *aromatic* compound. *(2 lines)* (1 mark)

(h) Ethanol is used as a biofuel to partially or completely replace

hydrocarbons in petrol. Hydrogen is another alternative fuel.

(i) Give one way in which ethanol is a more sustainable fuel than hydrocarbons in petrol. *(2 lines)* (1 mark)

(ii) Give one way in which the burning of hydrogen instead of ethanol is better for the environment. *(2 lines)* (1 mark)

Total: 34 marks

■ ■ ■

Candidates' answers to Question 4

Candidate A

(a) (i)

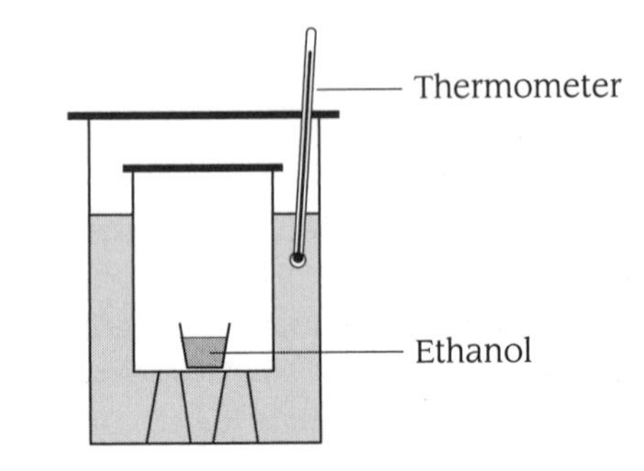

Candidate B

(a) (i)

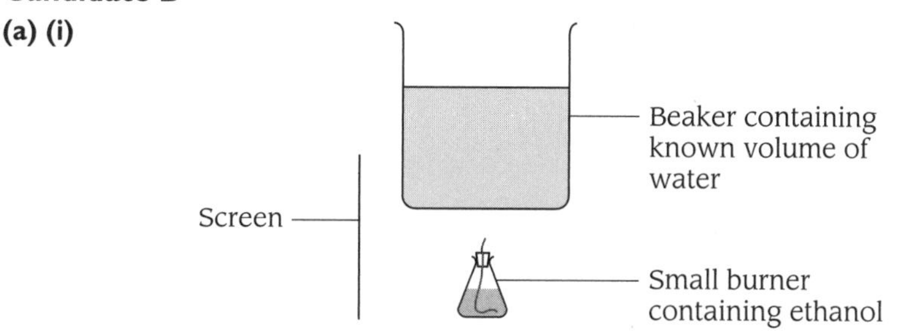

Neither scores full marks. Candidate A has been confused by the memory of a bomb calorimeter. This apparatus would not work well, but Candidate A would score 1 mark out of 3 for showing the thermometer in the water. This is the mark that Candidate B loses, as he has left this out. The rest of the diagram is fine, so he would score 2 out of 3 marks.

Candidate A

(a) (ii) Energy = $2.03 \times 4.2 \times 20 = 0.17$ (J)

Candidate B

(a) (ii) Energy given out = $500 \times 4.2 \times 20 = 42\,000$ (J)

Candidate A has not realised that it is the water which is being heated; thus the 'mass' referred to is the mass of the water, not the mass of the fuel. This makes her answer wrong from the start. She has also converted to kilojoules, which is incorrect as the answer was required in joules. She does not score.

Candidate B has got it right and scores the mark.

Candidate A

(a) (iii) $0.17 \times 46/2.03 = 3.9$ (kJ mol^{-1})

Candidate B

(a) (iii)

$$\frac{42\,000}{1000} \times \frac{46}{2.03} = -951.72 \text{ (kJ mol}^{-1}\text{)}$$

Candidate A has recovered her stride and would score marks for the numerical answer on the 'error carried forward' principle. She has also written her answer to two significant figures. She has left out the minus sign, though, since it is the enthalpy change that is asked for and enthalpy change has a sign convention.

Candidate B has got the right method. He has converted J to kJ by dividing by 1000 and then scaled up from 2.03 g to 1 mole (46 g) by multiplying by 46/2.03. He has also remembered the minus sign. He has, however, given too many significant figures, so he would lose a mark. The correct answer is –950 (kJ mol^{-1}) to two significant figures. Both score 2 out of 3 marks.

question

Candidate A

(a) (iv) No heat losses

Candidate B

(a) (iv) No heat absorbed by the calorimeter

Both are correct and score the mark.

Candidate A

(a) (v) 0.01/2 = 0.5%; 0.5/20 = 2.5%

Candidate B

(a) (v) 0.02/2 = 1%; 0.1/20 = 5%

Candidate A scores 1 out of 2 marks. She has not noticed that there are two readings involved in each measurement, thus the uncertainty should be doubled. Candidate B has this correct and scores 2 marks. Notice that these uncertainties indicate that there is no point in using a more accurate balance unless a much more accurate thermometer is used.

Candidate A

(a) (vi) Heat is lost in the apparatus.

Candidate B

(a) (vi) Most of the water produced is vapour. The standard state for water is liquid.

Candidate A has not read the question carefully. Heat loss is an experimental inaccuracy, so will not score. Candidate B has got it right *and* given enough detail to score the mark.

Candidate A

(b) (i)

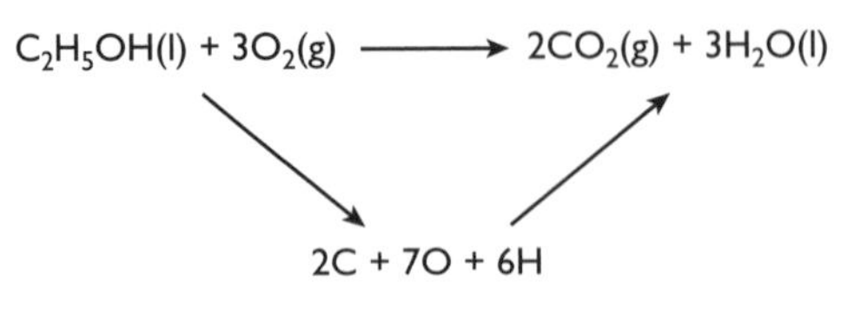

Candidate B

(b) (i)

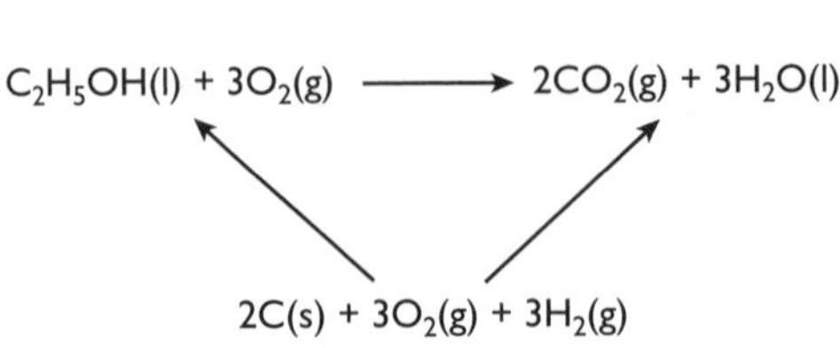

Candidate A has got the right idea of using the elements, but she has made a large number of errors. First, the enthalpy changes of formation required start from the elements *in their standard states*, not single atoms as shown. Also, there are no state symbols. So Candidate A is awarded 1 mark for the right 'shape' of the cycle. (The fact that the left-hand arrow is in the wrong direction will be considered in the mark scheme for the next part.)

Candidate B has nearly got it right. There is just one error — it should be $3.5O_2$, as there are seven oxygen atoms on each side, not six. So Candidate B would get 1 out of 2 marks, the 'shape' and the state symbols (at 298 K, as specified in the question) being correct.

Candidate A

(b) (ii) Answer = 394 + 286 − 277 = 403 kJ mol^{-1}.

Candidate B

(b) (ii) $\Delta H^{\ominus}_{c,298}$ = (2 × −394) + (3 × −286) − (−277) = −1369 kJ mol^{-1}.

Candidate A has again made lots of mistakes. First, she has failed to multiply the $\Delta H^{\ominus}_{f,298}$ values for CO_2 and H_2O by two and three respectively. Second, she has the signs round the wrong way. Even so, she has calculated the answer correctly from her values and has given the correct units. Does she get the third mark, therefore, with error carried forward? Unfortunately not, as she has failed to put in the plus sign. *Remember, ΔH values must have a sign, either plus or minus.* Of course, she ought to have realised that all enthalpy changes of combustion are negative and gone back and checked her work.

Candidate B has got the correct answer and scores 3 marks. He has multiplied the enthalpy changes of formation of CO_2 and H_2O by their respective number of moles and added them. He has then subtracted the enthalpy change of formation of ethanol, keeping a careful eye on the signs.

His earlier mistake with the oxygen does not matter, as the $\Delta H^{\ominus}_{f,298}$ value for O_2 is zero (as it is for all elements in their standard states).

Candidate A

(c)

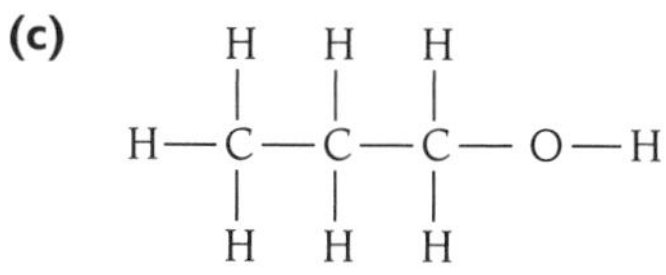

Candidate B

(c) $CH_3CH_2CH_2OH$

Candidate A is correct and scores 1 mark. Candidate B has given a structural formula but not the full structural formula, which shows all the bonds and atoms. Therefore he does not score.

Candidate A

(d) (i) 3 × 24 = 72 dm^3

Candidate B

(d) (i) 3.0 dm^3

Candidate A has become confused and worked out the volume of three moles of gas — not what is required here. (Note that the 24 dm^3 would always be given if you needed to use it.) She does not score.

Candidate B has seen that there is a 3:1 ratio between moles of oxygen and moles of ethanol vapour. Since all gas molecules occupy approximately the same volume (at the same temperature and pressure), all that is required is to multiply one by three. He scores 1 mark.

Candidate A

(d) (ii) $\frac{2.3}{46} \times 24 = 1.2\ \text{dm}^3$

Candidate B

(d) (ii) Moles ethanol = 2.3/46 = 0.05

Moles CO_2 = 0.10 = 4.4 g

Candidate A has not made it very clear what she is doing and she has got the answer wrong, so should she score any marks? Here the mark scheme tells examiners to look for '2.3 × 18/46' to reward in the right context, so she would score 1 mark. Her only mistake is to forget that one mole of ethanol burns to form two moles of carbon dioxide, from Equation 4.1. However, you may think this is generous and, I hope, learn to put in more working.

Candidate B has started well, but has lost track halfway through. He scores 1 mark for correctly calculating the moles of carbon dioxide, but then he calculates the *mass* of CO_2, rather than its volume. This is where correct chemistry does not score, as it does not answer the question. He scores 1 mark. The correct answer is 2.4 dm^3.

Notice that the examiner has been kind here and given the M_r of ethanol. You could also be asked to calculate this from the A_r values from the periodic table on the *Data Sheet.*

Candidate A

(e) (i) There are more ways of arranging the molecules of a gas than there are of a liquid.

Candidate B

(e) (i) Entropy is disorder. Liquid ethanol has less disorder than gaseous oxygen.

Candidate A has given the better answer and scores 2 marks. The question asks for an explanation and she has given it. It would have been safer to have mentioned ethanol and oxygen, the specific examples, but there is enough here to score both marks. The crucial points are 'ways of arrangement' and 'molecules'.

Candidate B's answer is weaker. The first sentence is an alternative way of scoring 1 mark, but the second virtually re-states the question and does not add any new information, such as 'there are more ways of arranging the particles in a gas than in a liquid'. He scores 1 mark.

Candidate A

(e) (ii) Increase, as there are more molecules.

Candidate B

(e) (ii) There are more gas molecules on the right-hand side of the equation, so the entropy will increase.

Candidate A has been vague again — more molecules where? She scores 1 mark for 'increase'.

Candidate B has given a full and correct answer, and scores 2 marks. Note that here the reason needed is for the answer given, not an answer in terms of ways of arranging the molecules.

Candidate A

(f) (i)

```
  H  H
  ×• ×•
H•×C•×C•×O•×H
  ×• ×•
  H  H
```

Candidate B

(f) (i)

```
  H  H
  ×• ×• ••
H•×C•×C•×O•×H
  ×• ×• ••
  H  H
```

Candidate A has nearly got it right, but she has forgotten that oxygen has six electrons in its outer shell (as it is in group 6) and so it has two *lone pairs* which must be shown. She does not score. Candidate B has drawn the correct diagram, and scores 1 mark.

Candidate A

(f) (ii) It does not show the shape.

Candidate B

(f) (ii) It does not show the electron charge-clouds.

Both are correct and score the mark.

Candidate A

(f) (iii) 1. 109 2. 180

Candidate B

(f) (iii) 1. 109° 2. 109°

Candidate A's answer to part (i) is right, if we excuse her for not putting the degree sign (which examiners usually do — be careful, though) Part (ii) is wrong and possibly follows from the omission of the lone pairs in the dot–cross diagram. She scores 1 mark.

Candidate B has them both right, and scores 2 marks.

Candidate A

(f) (iv) The pairs of electrons repel each other.

Candidate B

(f) (iv) There are four groups of electrons (two bonding pairs and two lone pairs) which repel each other and maximise the repulsion.

Candidate A is vague in not specifying the number of groups of electrons, but is right to state that they repel, so she scores 1 mark. However, once again, she has given an imprecise answer. If she had continued to say '...and get as far away from each other as possible', she would have scored another mark.

Candidate B is well on track, but the electrons get as far apart as possible to *minimise* repulsion. He therefore fails to score the third mark, though he scores the other two.

Candidate A

(g) Aliphatic, because it has no double bonds.

Candidate B

(g) Aliphatic, because it does not have a ring.

Both are correct to state that it is aliphatic, but both score zero. Examiners cannot award marks for a 50/50 choice, so the 1 mark is for the reason, once the correct answer has been chosen. There are many unsaturated compounds that are aliphatic, and there are some cyclic (ring) compounds (like cyclohexane) that are not aromatic. Aromatic compounds contain a *benzene ring*; aliphatic compounds do not.

Candidate A

(h) (i) It can be grown, so it won't run out.

Candidate B

(h) (i) The carbon dioxide it gives off when it burns is what it absorbed when it grew.

Candidate A is brief, which is unwise. Ethanol can be *made from things (such as cereal crops)* that are grown. This is close enough to score 1 mark, but remember to give as much detail as you can. Candidate B has left out that step too, and has gone in a different direction, which (while correct) does not answer the question about sustainable fuels, so will not score.

Candidate A

(h) (ii) Hydrogen gives no CO_2.

Candidate B

(h) (ii) Hydrogen burns to water only.

Both are correct and score the mark, though Candidate B would have been wise to add the point that ethanol produces CO_2 as well.

Candidate A has scored 16 marks and Candidate B 25 marks.